Let's Build Better, Safe. Kind And Goal-Oriented World, Not Just For People But For All Living Things.

The Comprehensive Guide On Corporate Social Responsibility

For

"PROFESSIONALS

&

CONSULTANTS"

Mr. Nitesh K Jadhav

The Comprehensive Guide On CSR For Consultant

   The Comprehensive Guide On CSR For Consultant

The Comprehensive Guide On CSR For Consultant

The Comprehensive Guide On CSR For Consultant

In my six-plus years as a professional, I've learned how important it is to have a guidebook on corporate social responsibility. As I've worked on different jobs and projects, I've come to realise that many people who work in social development sector want to move up in their careers but don't know how. I worked hard to make this book, which will help you advance your career in the field of corporate social responsibility. I really hope you enjoy reading this book and that it helps you do well as a CSR specialists.

The goal of making a handbook is to give people who work in social development and corporate social responsibility a step-by-step guide to CSR consulting. This manual was made to help everyone who works in social development and corporate social responsibility become a fully trained sustainability and corporate social responsibility consultant. This book is for people who want to start their own nonprofit organisation and are interested in working in the field of social development. It gives a step-by-step guide on how to do this.

There are hundreds of corporate social responsibility consulting companies around the world that help thousands of businesses with their corporate social responsibility and sustainability programmes. All of these companies hire very smart people who help shape the future of people, the environment, and the business. Their actions, goals, and ways of getting things done are important for making society better. This business and consulting groups are always looking for new people with fresh ideas. They are also looking for highly skilled people to work with them on

CSR and sustainability projects that will have the most good effects possible.

This handbook is meant to give context for dealing with day-to-day CSR operations, as well as for making CSR plan and putting it all into action. This Handbook is meant to fill the gap between broad, high-level overviews and clear, specific instructions that are made to fit the needs of each operation. The Handbook points to other tools and helps come up with policies and practises that are best for CSR.

     The Comprehensive Guide On CSR For Consultant

In Gratitude

As I write this, my heart is filled with enthusiasm and gratitude for the assistance, inspiration, and advice you provided me while writing my book. To my devoted family members Aai, Anna, Dear Bayko Yogita, Dear Brother Sachin, and My son Rudhvik, as well as to my life mentor Mr Rohidas K Jadhav (BE. Manufacturing Engineering (BIT's Pilani) Bharat Forge), I would like to express my gratitude for your support and understanding. My wonderful friends Ajinkya, Sunil, Priya, Abhijit, Suresh, Vijay, Mahesh, and Ritesh, whose support never ceased and whose ears were always willing to listen, gave me the determination to overcome my self-doubt and writer's block. My dedicated teachers, Ms. Ujjawala Masdekar Ma'am, Ms Nagmani Ma'am, Ms Anjali Maydeo Ma'am, Dr Deepak Walawkar Sir, Dr. Mahesh Thakur sir, Ms Sharmila Ramteke Ma'am, Ms Neha Sathe Ma'am, Dr Dada Dadas sir, Ms Deepa Tak. My Professional Superiors and Mentors Mr. Ermino (WB Safegaurd Team) Mr. Sagar Kabra, Mr. Shinnoy Mathew, Ms. Sarita Bahal, Mr. Chayan Pardi, Mr. Rupesh Pawar, Ms. Madhuri Sanas, Mr. Prashant Bamb (MLA), Mr. Anil Ghuge, Ms. Anjana Goswami, Mr. Pravin Katke, Mr Paulo Nota (Mozambique-Africa) Your knowledge, insight, and faith in my mission have been of great assistance to me.

The Karve Institute of Social Social Service in Pune is where I got my Master's. I really appreciate all of my teachers, Mentors & Classmates from 2015-2017.

 The Comprehensive Guide On CSR For Consultant

Extraordinary Thankfulness

I will always be grateful to "Techknowgreen Solutions Limited," Dr. Ajay Ojha, Mr. Prasad Pawer, Mr. Nadeem Sir, Mr. Vinay Bedekar, and all of TSL's other employees for their constant help as I wrote my book. Your huge help, resources, and enthusiasm for my idea were crucial to me reaching my goal. I can't tell you how much your dedication to writing and imagination has changed my life, but I'm grateful for the new chances you've given me because of it. I can't wait to show you the finished work, and I hope it makes you as happy as it has made me. We really value how much you added to this amazing experience.

Why Handbook for CSR Specialists

A Corporate Social Responsibility (CSR) consultant handbook is a complete guide or manual that tells CSR experts what they need to know, how to do things, and what the best practises are. Its goal is to help CSR consultants, experts, social workers, and counsellors, as well as companies, think about social and environmental issues when making decisions and running their businesses.

The following are some benefits of a CSR consultant handbook:

Guidance and structure: The handbook gives CSR experts a base for understanding CSR ideas, principles, and practises. It talks about how to evaluate a company's current CSR procedures, find places where they could be better, and make plans for putting CSR projects into action.

Best Practises: The handbook usually has case studies and examples of successful CSR programmes from a wide range of fields. CSR consultants can learn from and get ideas from these cases, which help them understand how multinational companies have successfully put CSR programmes into place.

Tools and Resources: A consultant's handbook will often have tools, models, and resources that they can use in their work. Some of these tools are assessment checklists, ways to engage stakeholders, measurement structures, and requirements for reporting.

 The Comprehensive Guide On CSR For Consultant

Insights about a certain business: Each industry has its own set of CSR challenges and opportunities. The handbook may have information about important challenges and trends in a number of different areas. With this knowledge, consultants can make sure that their suggestions and plans fit the needs of each sector.

Strategies for Engaging Stakeholders: Engaging stakeholders is an important part of CSR, and the guide can show how to connect well with both internal and external stakeholders. It could talk about things like stakeholder mapping, communication strategies, and making alliances with other groups. All of these things are important for the success of CSR activities.

The updated landscape of CSR : is always changing as new trends, rules, and stakeholder expectations come out. A consultant handbook could keep consultants up-to-date on the latest developments in CSR, giving them the skills and knowledge they need to deal with both current and future problems and concerns.

Professional Development: CSR consultants can use the guidebook as a tool for professional development to improve their skills, learn more, and stay up to date on industry norms. It can also link to books, articles, and training courses that can help you learn more.

Impact Monitoring and Reporting: It is important to keep track of and report on the effects of CSR activities in order to figure out how much they are worth and how well they work. The handbook could have suggestions for defining key performance indicators (KPIs), setting up

   The Comprehensive Guide On CSR For Consultant

methods for monitoring and evaluating, and telling stakeholders about the results of CSR.

Ethical Considerations: CSR experts often find themselves in tough moral situations at work. The manual can help you deal with these problems and make moral choices. It could talk about things like the rights of stakeholders, social and environmental justice, and the right way for a business to act.

Sustainability Reporting and Communication: The handbook shows experts how to do reporting and communication about sustainability. It goes into detail about the most important reporting structures, data collection methods, and ways to tell internal and external stakeholders about CSR goals and successes.

Overall, a CSR consultant handbook is a very important tool for professionals in the field. It gives them the knowledge, advice, and tools they need to help organisations with their CSR strategies and activities.

   The Comprehensive Guide On CSR For Consultant

CHAPTER 1: THE CONSULTANT

A consultant is a trained professional who helps people, groups, or companies in a certain area by giving them advice and direction. Consultants are often hired to give clients particular knowledge, skills, and insights that can help them solve problems, make better decisions, and improve their performance or efficiency in a certain area. Consultants are needed for a wide range of reasons that companies and people face in different situations. Here is a list of some of the most important reasons to consult: Consultants are experts in their areas and have a wide range of skills. They know things that are hard to find among the company's employees or in the organisation as a whole. Consultants give a fair and neutral point of view. Their lack of bias or internal politics can help them see problems, give insights, and suggest answers that are free of politics or biases. Organisations may not have enough time, money, or skills to handle certain jobs or problems well.

Consultants can take over and focus only on the problem at hand, which leads to good answers. Consultants work with a wide variety of industries and clients, which helps them learn about best practises and successful strategies that can be used in each client's unique situation. Consultants can be hired for short-term projects. This lets companies increase or decrease their staff based on their current needs without having to hire permanent workers. When only a small amount of knowledge is needed for a short time, it may be

    The Comprehensive Guide On CSR For Consultant

cheaper to hire consultants than to keep an in-house team working full-time. During times of change or transformation in a company, consultants can help guide people through problems and help them find solutions that work.

Consultants can help find and reduce project risks, making sure the project goes more smoothly and reducing the chance of making mistakes that cost a lot of money. Consultants often help companies train and develop their workers, giving them new knowledge and skills and helping them grow and get better over time. Consultants can improve an organization's general performance, productivity, and efficiency by giving advice on how to improve processes and optimise workflows. By researching and analysing the market, consultants can help businesses make smart choices and stay competitive in a business world that is always changing. Some projects or initiatives may need very specialised knowledge that isn't easy to find within the company. In this case, hiring consultants may be necessary.

Consultants can work in many different fields, such as management, finance, human resources, marketing, technology, healthcare, education, and many more. They might work for consulting firms, be self-employed, or hold positions in bigger organisations.

The key characteristics of a consultant include:

Expertise	Consultants are experts in their fields and have a lot of knowledge and experience that is important to the problems or challenges that their clients are facing.

   The Comprehensive Guide On CSR For Consultant

Advisor	As outside advisors, consultants offer an unbiased and objective point of view, which can be helpful when looking for chances and making suggestions for how to deal with them.
Problem-Solving	Consultants are good at looking at complicated situations, figuring out what the problems are, and coming up with good ways to solve them or reach certain goals.
Customized Solutions	They make sure that their suggestions fit the wants and circumstances of each client by taking into account things like the client's industry, size, budget, and organizational culture.
Project-based work	Consultants are often hired for specific projects or jobs with clear goals and deadlines. When the job is done, they may move on to work with other clients or on other projects.
Communication	Consultants need to be able to talk to their clients in a way that makes their ideas, results, and suggestions clear and convincing.
Trends	Good consultants keep up with the latest research, trends, and developments in their areas so they can give the most relevant and up-to-date advice.

Consultants are very important to organisations because they help them solve problems, improve performance, and reach their goals by using their knowledge and insights in a professional and collaborative way.

1.1 Consultant For Corporate Social Responsibility

A Corporate Social Responsibility (CSR) expert is a professional who advises businesses and organisations on how to run their businesses in a way that is both sustainable and socially responsible. These consultants help businesses use CSR principles in their strategies, operations, and culture to make a positive effect on society and the environment while also ensuring long-term business success.

A CSR consultant usually knows a lot about things like sustainability, environmental management, social effect, ethical business practises, stakeholder engagement, and reporting. They work closely with businesses to create CSR strategies that match the organization's goals, values, and the expectations of its stakeholders. CSR consultants also help companies put in place specific programmes and initiatives to deal with environmental, social, and governance (ESG) problems, get employees more involved, and improve the company's reputation.

The role of a CSR consultant can change based on what the client wants and how big the project is. They may work on their own, for a consulting firm, or as part of a bigger CSR team within a company. Their ultimate goal is to help businesses make decisions that are more responsible and sustainable, which will have a good effect on society and the planet.

    The Comprehensive Guide On CSR For Consultant

Corporations should think about getting a Corporate Social Responsibility (CSR) consultant for several good reasons:

Sustainable Business Practises	CSR consultants know how to incorporate sustainability concepts into a company's operations, supply chain, and product development. For example, a consultant could help a manufacturing business find ways to use less energy, make less trash, and adopt circular economy practises. This would reduce the company's impact on the environment.
CSR Strategy Development	Consultants help companies come up with a CSR strategy that fits with their values and goals. For example, a consultant might help a retail company come up with a CSR plan that includes projects to help local communities, promote sustainable sourcing, and reduce plastic waste.
Environmental Management	Consultants know about environmental problems and can tell businesses how to deal with them in a responsible way. For example, a consultant might help a construction company use eco-friendly building methods to reduce the damage to local environments and keep natural resources from running out.
Social Impact Assessment	CSR consultants can do social impact assessments to find out what kind of effects a company's actions might

 The Comprehensive Guide On CSR For Consultant

	have on local communities, both good and bad. Then, they might suggest ways to maximise the good results and minimise the bad ones. For instance, a consultant might work with a mining business to make sure that their operations help the local community by creating jobs and improving infrastructure while causing the least amount of people to move or damage to the environment as possible.
Ethical Business Practises	Consultants can help companies make and use codes of behaviour and ethical business rules. For example, a consultant could help a financial company make sure it has fair lending practises, clear financial reporting, and smart investments.
Stakeholder Engagement	CSR experts are good at figuring out who the key stakeholders are and what they want and need. They can help a company and its stakeholders have important conversations that build trust and understanding for both sides. For example, a consultant could help a tech company talk to NGOs and community leaders about data privacy and cybersecurity issues.
Reporting and Standards	Consultants are familiar with global CSR reporting structures and standards, such as the National Voluntary Guidelines on Social, Environmental, and Economic Responsibilities of Business, the

The Comprehensive Guide On CSR For Consultant

	Global Reporting Initiative (GRI), or the United Nations Sustainable Development Goals (SDGs). They can help companies make CSR reports that are honest and clear about how they are doing in terms of social and environmental responsibility.
Strategic Alignment	A CSR consultant can help align CSR projects with the business plan of the company as a whole. This makes sure that the organization's CSR efforts are aligned with its purpose, values, and long-term goals, which makes them more effective and influential.
Stakeholder Engagement	Consultants can find and work with key stakeholders, such as customers, workers, investors, communities, and non-governmental organisations (NGOs). By knowing what they want and what worries them, businesses can build stronger relationships, improve their reputation, and gain more trust.
Risk Mitigation	Companies can reduce environmental, social, and governance (ESG) risks by using CSR practises. A CSR expert can find possible risks and give advice on how to manage and lessen them.
Brand name	If a company's CSR efforts are successful, they can improve the brand's name and image. Consumers and investors often prefer to support businesses that care about the

 The Comprehensive Guide On CSR For Consultant

	community. This can lead to more loyal customers and investors who trust the company more.
Innovation and Efficiency	CSR consultants can help the organisation be more innovative, which can lead to the development of business practises, goods, and services that are more sustainable and efficient.
Competitive Advantage	Businesses can gain a competitive edge in the market by incorporating CSR into their business models. Being a responsible and environmentally friendly business can help you get customers and employees who value ethical and socially conscious companies.
Positive Impact	By working with a CSR consultant, companies can make sure that their CSR efforts are well-designed, effective, and focused on making a positive addition to society and the environment.
Long-Term Sustainability	CSR experts can help companies come up with sustainable business practises that deal with current problems and think about the organization's long-term viability in a world that changes quickly.
Employee Engagement	Getting employees involved in important CSR projects can boost morale and get them more interested in their work. People are usually

 The Comprehensive Guide On CSR For Consultant

	proud to work for a company that cares about helping people and the earth.

A Corporate Social Responsibility (CSR) consultant has a wide range of skills and knowledge related to sustainable business practises, environmental stewardship, social impact, ethical considerations, and stakeholder involvement. In conclusion, hiring a CSR consultant can be a smart move for businesses. It can lead to a better brand image, better relationships with stakeholders, and long-term success, all while doing good things for society and the environment.

CHAPTER 2: HISTORY OF THE RISE OF INDUSTRY

The story of industrialization has been told in history books. Scholars have different ideas about when it started and when it finished, but it was mostly between 1760 and 1840.

The Industrial Revolution started in Great Britain, continental Europe, and the United States between 1760 and 1840. It was a time when the whole world's economy moved towards more efficient and stable ways of making things. This change included the switch from making things by hand to making things with machines, new ways of making chemicals and iron, more use of water and steam power, the development of machine tools, and the rise of the mechanised factory system. Since the textile industry was the first to use modern manufacturing methods, it quickly surpassed other industries in terms of jobs, output value, and capital invested.

Corporate social responsibility (CSR) and the past of industrialization are linked, with the social and environmental effects of industrialization influencing the development of CSR.

Early Industrialization: The Industrial Revolution in the 18th century caused a big change from a farming economy to an industrialised one. During this time, industrialization caused a lot of problems in society and the environment. Poor working conditions, child labour, pollution, and taking advantage of natural resources were common, which made people angry and led to calls for change.

As industrialization moved forward, trade unions and labour groups began to fight for workers' rights and better working conditions. These groups tried to fix social problems and make life better for workers. They pushed for fair treatment and safe working conditions, which led to early ideas about social responsibility.

Philanthropy and Welfare Movements: In the late 1800s and early 1900s, businessmen like J.RD. Tata, Andrew Carnegie, and John D. Rockefeller were known for their charitable work. They set up charities and gave a lot of money to education, health care, and building up the community. Even though these efforts were done for personal reasons, they helped build the foundation for the idea that businesses can help improve society.

Rise of Corporate Citizenship: In the middle of the 20th century, the idea of corporate citizenship became more popular. Scholars like Howard R. Bowen and E. Merrick

Dodd pushed the idea that companies have responsibilities that go beyond their financial goals and that they should think about how their actions affect society as a whole. This is where the idea of corporate social responsibility (CSR) came from.

Environmental Awareness and Regulation: As environmental awareness grew in the second half of the 20th century, environmental laws and standards were put in place. Pollution and the depletion of resources are both made worse by industrial operations. Because of this, businesses started to take environmental problems into account in their work, which is why environmental sustainability is now on the CSR agenda.

Globalisation and Pressure from Stakeholders: As globalisation and connectivity grew, groups like consumers, investors, workers, and civil society started to ask firms to be more open and take more responsibility. Businesses put CSR practises in place to protect their image and reduce risks. They did this because they were worried about labour practises, human rights, the environment, and community development.

As a Business Strategy, CSR: In the last few decades, CSR has changed from a side project to a key business strategy. Companies realised that doing things in a responsible way could help them be more competitive, attract customers, keep good employees, manage risks, and build long-term value. CSR is generally seen as an important part of a company's strategy. It includes things like taking care of the environment, doing business in an honest way, helping the community, and managing the supply chain in a responsible way.

     The Comprehensive Guide On CSR For Consultant

The Industrial Revolution was an important changing point in history. The Industrial Revolution changed almost every part of daily life in terms of material growth. The average income and population started to go up in a way that had never happened before. As a result of the Industrial Revolution, for the first time in history, the general standard of living in the Western world began to rise gradually.

Since the end of the 18th century, the rise of modern industry has led to a huge increase in urbanisation and the rise of new great cities, first in Europe and then in other places, as people moved from rural areas to cities to take advantage of new opportunities. In 1800, only 3% of the world's people lived in towns. By the beginning of the 21st century, that number had grown to almost 50%.

The economic revolution of the 21st century will be shaped by new technologies, especially automation. At the start of the 20th century, groups that didn't make money started to form. Nonprofits and businesses have started to work on a wide range of social problems, such as human rights, equality, poverty, the environment, and so on.

   The Comprehensive Guide On CSR For Consultant

CHAPTER 3: HISTORY OF SOCIAL RESPONSIBILITY IN BUSINESS

The Industrial Revolution is where business social responsibility got its start. Companies became the centre of society after the Industrial Revolution. Farmers moved to cities and company towns to work in workshops where they could make more money. Early businessmen saw the need for companies to care about their communities. Businesses that did well knew that a healthy staff made them more productive and brought in more money. Others became interested in helping people in need and giving back because it made them feel good. Corporate social responsibility, also called corporate sustainability, sustainable business, corporate citizenship, social effect, conscious capitalism, and responsible business, has been a focus of many businesses and stakeholders since the 1960s.

We can't find many cases of social responsibility on the part of businesses in the 1800s.

Jamsetji Tata : Jamsetji Tata, who started the Tata Group, is often thought of as one of India's most generous people. He started places like the Indian Institute of Science and the Tata Institute of Fundamental Research that teach and do research. He helped with things like health care, social welfare, and building up the neighbourhood.

Sir Dorabji Tata Trust : The Sir Dorabji Tata Trust was started in 1932 by Sir Dorabji Tata, the oldest son of Jamsetji Tata, who started the Tata Group. Sir Dorabji Tata is also known as Sir Dorabji. It is one of India's oldest organisations that helps people without regard to religion. Sir Dorabji Tata died in 1932 and left the trust all of his money. The Sir Dorabaji Tata Trust carries on Jamsetji Tata's philanthropic legacy by starting India's first institutions like the Tata Institute of Social Science in 1936, the Tata Memorial Centre for Cancer Research and Treatment in 1941, and the Tata Institute of Fundamental Research in 1945. The legacy of philanthropy and corporate social responsibility goes on and on.

John D. Rockefeller: John D. Rockefeller was a well-known benefactor in the 20th century. John D. Rockefeller started the Standard Oil Company. He became one of the richest people in the world and a very important charity. In 1884, Rockefeller gave a lot of money to the Atlanta Baptist Female Seminary, which was for black women and later became Spelman College. Laura Spelman Rockefeller, who was his wife, worked for civil rights and equal rights for women. John and Laura gave money to the Atlanta Baptist Female Seminary. Spelman College is an all-

   The Comprehensive Guide On CSR For Consultant

women's historically black college or university in Atlanta, Georgia. It is named after Laura's family.

Andrew Carnegie: Andrew Carnegie, a Scottish-American businessman and humanitarian, was one of the most generous people of the late 1800s and early 1900s. He gave a big chunk of his money to schools, educational institutions, and groups that work for peace and study.

Ratan Tata: Ratan Tata, a former chairman of the Tata Group, has taken part in a number of charitable endeavours. He has contributed to causes linked to healthcare, rural development, education, and disaster relief through the Tata Trusts. Many people's lives have been impacted by Ratan Tata's generosity, which has also helped to promote nation-building.

Bill and Melinda Gates: Since the late 20th century, Bill Gates and his ex-wife Melinda Gates have been known for their charitable work. Through the Bill & Melinda Gates Foundation, they have focused on world health, education, reducing poverty, and access to technology. They have spent a lot of money and made a lot of progress in these areas.

Warren Buffett is an American business owner and investor who is known for the good things he does for people. He gave a lot of his money to the Bill and Melinda Gates Foundation and other charities, with a focus on education, reducing poverty, and health care.

Azim Premji: Azim Premji, the father of Wipro Limited, is one of the most well-known people in modern India who has done good things for others. He has done a lot for education through the Azim Premji Foundation, which

 The Comprehensive Guide On CSR For Consultant

focuses on improving basic education and training teachers. Through his charity work, he tries to meet the educational needs of kids in need all over the country.

George Soros: George Soros, a Hungarian-American businessman and philanthropist, has been interested in giving back for a long time. Soros's Open Society Foundations have helped people all over the world work for free government, human rights, education, and social justice.

Milton Hershey : Milton Hershey opened his chocolate company in 1894. He made a park for his company town and made sure that families who worked in the factory had fun things to do. The town also had a community greenhouse and plant nursery, and Hershey workers were able to buy homes with low-interest loans. Hershey and his wife Catherine started the Milton Hershey School for Orphans in 1909. The private school still teaches nearly 2,000 kids who are in need.

Marta Abreu de Estévez : was a well-known woman in the centre of Cuba, especially in the city and province where she was born, Santa Clara. She was known as "the great benefactor" because she helped poor people all the time, gave money to the city, and backed the freedom war. Marta Abreu was rich, so she was able to travel to Europe and the United States at a young age. There, she met important historical figures and learned about the differences between Cuba and the most developed countries, as well as the struggles Cubans face, especially in the country's small rural cities and towns.

   The Comprehensive Guide On CSR For Consultant

Mukesh Ambani & Neeta Ambani, The chairman and major shareholder of Reliance Industries Limited has been involved in a number of charitable projects. Even though he is best known for his work in business and the professional world, he has also done a lot to help people in need. In 2010, Reliance Industries started giving back to the community by making the Reliance Foundation. The foundation works on a wide range of issues related to helping people, such as healthcare, education, rural development, and disaster aid. It does a number of things to improve neighbourhoods and make the lives of poor people better.

Oprah Winfrey is an American television mogul and philanthropist who has given a lot of money to many different groups. She has helped with, among other things, education, women's rights, health care, and disaster aid.

Mark Zuckerberg and Priscilla Chan: Mark Zuckerberg, who helped start Facebook, and his wife, Priscilla Chan, have become big helpers in recent years. Through the Chan Zuckerberg Initiative, they want to address social problems by improving education, scientific study, and community development.

Shiv Nadar: Shiv Nadar, who co founded HCL Technologies, has done a lot of charity work through the Shiv Nadar Foundation. The organisation works on education with programmes like the Shiv Nadar University and VidyaGyan, a leadership academy for students in rural areas. Nadar's kindness helps both rural areas and people who need medical care.

   The Comprehensive Guide On CSR For Consultant

Kiran Mazumdar-Shaw: Kiran Mazumdar-Shaw, who started Biocon Limited, has done a lot to improve health care and schooling in India. She has worked to improve the infrastructure of health care, support research and development, and make sure that people can get reasonable health care. Public health projects in India have been helped by Mazumdar-Shaw's charity for a long time.

Tonny Robbins: Tony Robbins is known as a motivational speaker, a life guide, and an author. He has also done a lot for charity. His main focus has been on helping people improve themselves and gain power, but he has also given money to many good causes over the years. Robbins started the Anthony Robbins Foundation in 1991 to help change people's lives, their families' lives, and their neighbourhoods. The goal of the foundation is to give young people more power, feed, house, and educate homeless people, and help organisations that work in areas like youth leadership, prison reform, and foreign humanitarian work.

Henry welcome, Howard Hughes, Hans Wilsdorf, JK Lilly Sr, Edsel Ford, Lakshmi Mittal (ArcelorMittal),Adi Godrej (Godrej Group), Cyrus Poonawalla (Serum Institute of India), Pallonji Mistry (Shapoorji Pallonji Group), Yusuf Hamied (Cipla), K. Anji Reddy (Dr. Reddy's Laboratories), B.R. Shetty (NMC Health), Anand Mahindra (Mahindra Group), Dilip Shanghvi (Sun Pharmaceuticals), Vijay Shekhar Sharma (Paytm), Subhash Chandra (Essel Group), M.A. Yusuff Ali (Lulu Group), Savitri Jindal (Jindal Group), P.N.C. Menon (Sobha Limited), Baba Kalyani (Bharat Forge), S.D. Shibulal (Infosys), Brijmohan Lall Munjal (Hero MotoCorp), Shiv Nadar (SSN Institutions),

   The Comprehensive Guide On CSR For Consultant

Rahul Bajaj (Bajaj Group), Shri Prakash Lohia (Indorama Corporation), Ajay Piramal (Piramal Group), Anil Agarwal (Vedanta Resources), Venu Srinivasan (TVS Motor Company), Sunil Bharti Mittal (Bharti Enterprises), Pawan Munjal (Hero MotoCorp), Kumar Birla (Aditya Birla Group), Brij Mohan Khaitan (Eveready Industries India Ltd), Rajendra Pawar (NIIT Group), Sudha Murty (Infosys Foundation), Gautam Adani (Adani Group), N.R. Narayana Murthy (Infosys), Pankaj Patel (Cadila Healthcare), Swati Piramal (Piramal Enterprises), S. Gopalakrishnan (Infosys), Radhakishan Damani (DMart), P.R.S. Oberoi (Oberoi Group), Bhavish Aggarwal (Ola Cabs), Vineet Nayar (HCL Technologies), Venu Srinivasan (TVS Group), Vinod Khosla (Khosla Ventures), Vikram Lal (Eicher Motors), Malvinder Mohan Singh (Fortis Healthcare), Ajay Singh (SpiceJet), these are the few greatest philanthropists of the twentieth and twenty-first centuries. All of these are big business owners who have achieved the pinnacle of financial and social success in their lives.

Please keep in mind that this is not a complete list, and there have been many additional Indian philanthropists who have made substantial contributions to society. The philanthropic environment is wide and vibrant, with many people actively involved in humanitarian issues and making a difference in a variety of fields.

Corporate social responsibility (CSR) is a form of international private business self-regulation that aims to support charitable, activist, or philanthropic societal goals through community development, allocating financial grants to nonprofit organizations for the public good, and

   The Comprehensive Guide On CSR For Consultant

engaging in morally upright business and investment practices.

CHAPTER 4: THE GLOBAL CONTEXT OF CORPORATE SOCIAL RESPONSIBILITY

Global corporate social responsibility, or CSR, is the use of ethical and responsible business practises to solve social, environmental, and economic problems around the world. It means that companies need to realise they have a bigger impact than just their financial success and work actively on a global scale towards sustainable development, positive social change, and environmental stewardship.

Global CSR is based on the idea that companies have a duty to society and the environment. It goes beyond normal business operations by trying to include social and environmental issues in a company's basic strategies and practises. Global CSR includes things like human rights, wage standards, community development, protecting the environment, and doing business in an honest way.

It recognises that companies aren't just there to make money; they also have a duty to contribute to sustainable development. At the global level, CSR looks at how multinational companies (MNCs) affect a wide range of stakeholders, such as workers, communities, consumers, and the environment. It also tries to incorporate social and environmental problems into business plans and operations.

CSR is a global idea that is based on a few key beliefs and practises:

Ethical Business Conduct	Global CSR stresses how important it is for MNCs to run their businesses in a way that is morally right, open, and respectful of human rights. This means upholding human rights norms that are generally accepted, promoting ethical hiring practises, and fighting corruption.
Sustainability in the environment	Multinational corporations (MNCs) are required to reduce their impact on the environment and support sustainable practises all over the world where they do business. This means reducing greenhouse gas emissions, protecting natural resources, using good waste management methods, and helping with

   The Comprehensive Guide On CSR For Consultant

	campaigns to support renewable energy and slow down the effects of climate change.
Stakeholder Engagement	A key part of global CSR is working with stakeholders and getting along with them. MNCs should work with communities, NGOs, governments, and foreign organisations to make sure that what they do fits with what the public wants and needs. Participatory decision-making, community development projects, and working with organisations from civil society can all be put into this group.
Supply Chain Management	Multinational corporations (MNCs) need to make sure that the ways they do business in their supply networks are ethical and good for the environment. This means supporting ethical labour practises, promoting ethical sourcing, and keeping an eye on how suppliers treat the earth and people.
Collaboration and Advocacy	Multinational corporations (MNCs) can add to global CSR by working with other stakeholders like governments, non-governmental organisations (NGOs), and industry groups. This cooperation could lead to the creation of global standards, laws, and structures that support sustainable business practises and help solve global problems like climate change, poverty, and inequality.
Philanthropy and social	Global CSR often includes giving things and money to charitable causes

investments	and programmes. MNCs may pay programmes for social development, like health care, education, and reducing poverty, in order to have a positive effect on the places where they work.
Reporting and Transference	Multinational corporations (MNCs) are urged to report on their CSR efforts and their effects in a clear and honest way. This builds trust and lets stakeholders judge how well a company is doing in terms of its social and environmental impact. This makes the company responsible for its actions.

Global CSR is affected by a number of international norms and structures, such as the United Nations Global Compact, the Sustainable Development Goals (SDGs), and the OECD rules for Multinational Enterprises. These Structures set up rules and directions for how businesses should act responsibly, as well as a plan for how multinational companies can incorporate CSR into their operations around the world.

The United Nations Global Compact (UNGC) is a voluntary programme that encourages companies to adopt policies and practises that are good for the earth and people. It lists ten rules that companies should follow when it comes to human rights, work, the environment, and fighting corruption.

ISO 26000: This international standard gives advice on how to be socially responsible and helps businesses build CSR into their core business. It talks about a lot of different

　The Comprehensive Guide On CSR For Consultant

things, like how the company is run, how workers are treated, the environment, fair business practises, consumer issues, and how the community can be involved.

SDGs (Sustainable Development Goals): The Sustainable Development Goals (SDGs) were made by the United Nations and give the world a way to deal with important social, economic, and environmental issues. They are made up of 17 goals and 169 targets that deal with things like poverty, injustice, climate change, and making sure people aren't wasting too much. Businesses are asked to connect their goals for corporate social responsibility to the SDGs.

The Global Reporting Initiative (GRI) is a complete sustainability reporting system that helps organisations measure, manage, and communicate their economic, environmental, and social effects. It sets reporting standards for a number of CSR metrics, so companies can show how well they are doing.

OECD Guidelines for Multinational Enterprises: The OECD Guidelines for Multinational Enterprises are government-backed standards that tell businesses how to act responsibly in areas like employment, human rights, the environment, bribery, and consumer safety. They are meant to help multinational companies act in a way that is good for society and the earth.

Conventions of the International Labour Organisation (ILO): The ILO comes up with and supports international labour standards. These standards cover things like freedom of association, child labour, forced labour, and not discriminating against people. These conventions give companies a way to make sure that decent work conditions and fair labour practises are followed all along their global supply chains.

Project on Carbon Disclosure (CDP): The CDP is a global platform for sharing information that encourages companies to measure and report their environmental impacts, especially carbon emissions and risks linked to climate change. It gives organisations a standard way to report on their environmental success and helps them manage it.

Guiding Principles of the United Nations on Business and Human Rights: They spell out what companies should do to protect human rights, stop violations, and help people who have been hurt. The principles tell companies to do their "due diligence" to find, stop, and reduce bad effects on human rights.

4.1: The United Nations Global Compact (UNGC)

The United Nations Global Compact (UNGC) is a voluntary project that was started by the UN in 2000 to encourage businesses and organisations all over the world to adopt policies and practises that are good for people and the environment. It gives businesses a place to show that they are dedicated to making sure their operations and strategies are in line with ten globally accepted principles in the areas of human rights, labour, the environment, and fighting corruption.

The UN Global Compact is based on the idea that businesses, as important parts of society, have a responsibility to contribute to the greater goals of sustainable development and to solve important global

   The Comprehensive Guide On CSR For Consultant

problems. When a company joins the UNGC, it promises to protect and support the ten principles listed below:

Human Rights	<ul><li>Businesses should support and respect the protection of internationally proclaimed human rights.</li><li>They should ensure that they are not complicit in human rights abuses.</li></ul>
Labor	<ul><li>Businesses should uphold the freedom of association and the effective recognition of the right to collective bargaining.</li><li>They should support the elimination of all forms of forced and compulsory labor.</li><li>They should support the effective abolition of child labor.</li><li>They should eliminate discrimination in employment and occupation.</li></ul>
Environment	<ul><li>Businesses should support a precautionary approach to environmental challenges.</li><li>They should undertake initiatives to promote greater environmental responsibility.</li><li>They should encourage the development and diffusion of environmentally friendly technologies.</li></ul>
Anti-Corruption	<ul><li>Businesses should work against corruption in all its forms, including extortion and bribery.</li></ul>

   The Comprehensive Guide On CSR For Consultant

Companies that take part in the UNGC are encouraged to use these values in their operations, strategies, and company cultures. They have to make a public promise to follow the UNGC's principles, put out an annual Communication on Progress (COP) report that explains what they're doing, and take part in learning and dialogue activities to improve business ethics. On a platform set up by the UNGC, companies can share best practises, work with other partners, and reach their sustainable development goals. It wants to build a global network of companies that are committed to doing business in an honest way and that have a positive effect on society, the environment, and the economy. As of 2021, the UNGC has over 12,000 participants, including businesses, NGOs, labour organisations, and academic institutions. It is one of the biggest voluntary corporate sustainability projects in the world.

4.2 : The International Organisation for Standardisation (ISO): 26000

The International Organisation for Standardisation (ISO) made the international standard ISO 26000, which gives advice on social duty. It gives businesses of all sizes and types a clear way to add social responsibility into how they run and make decisions. ISO 26000 was made in 2010 after a process that included officials from governments, non-governmental organisations (NGOs), business, labour organisations, and consumer groups. ISO 26000 is a guidance standard that is meant to help companies understand and adopt socially responsible practises. Other

 The Comprehensive Guide On CSR For Consultant

ISO standards, on the other hand, set out certification requirements.

The standard is based on seven core subjects of social responsibility:

1. Organizational governance: This subject focuses on the organization's governance structure, ethical behavior, and accountability.
2. Human rights: It covers respect for internationally recognized human rights, including labor rights, non-discrimination, and supporting human rights within the organization's sphere of influence.
3. Labor practices: This subject addresses issues related to employment and the workplace, including fair employment practices, occupational health and safety, and employee development and training.
4. Environment: It emphasizes environmentally sustainable practices, resource usage, pollution prevention, and promoting environmental responsibility.
5. Fair operating practices: This subject focuses on ethical business practices, anti-corruption measures, and fostering fair competition.
6. Consumer issues: It highlights consumer rights, fair marketing and advertising, and promoting responsible consumption.
7. Community involvement and development: This subject encourages organizations to engage with and contribute to the communities in which they operate, supporting social and economic development.

   The Comprehensive Guide On CSR For Consultant

ISO 26000 gives organisations practical guidelines, best practises, and help on how to deal with each of these basic themes and include social responsibility in their plans, policies, and practises. It stresses how important it is for stakeholders to be involved, for things to be clear, and for things to keep getting better.

ISO 26000 is not a certification standard, but companies can use it to review and improve their social responsibility performance and let stakeholders know about their efforts. It is a great tool for companies that want to make sure their actions are in line with internationally accepted principles of social responsibility and contribute to sustainable development in a good way.

4.3: The Sustainable Development Goals (SDGs)

The United Nations adopted the Sustainable Development Goals (SDGs) in 2015 as a worldwide call to action to eradicate poverty, safeguard the environment, and promote peace and prosperity for all by 2030. The SDGs build on the success of the MDGs by addressing a broader variety of interconnected challenges, such as the social, economic, and environmental components of sustainable development.

The 17 SDGs encompass a wide range of objectives, each with specific targets to be achieved. The goals are as follows:

1. No Poverty
2. Zero Hunger
3. Good Health and Well-being
4. Quality Education
5. Gender Equality
6. Clean Water and Sanitation

     The Comprehensive Guide On CSR For Consultant

7. Affordable and Clean Energy
8. Decent Work and Economic Growth
9. Industry, Innovation, and Infrastructure
10. Reduced Inequalities
11. Sustainable Cities and Communities
12. Responsible Consumption and Production
13. Climate Action
14. Life Below Water
15. Life on Land
16. Peace, Justice, and Strong Institutions
17. Partnerships for the Goals

The Sustainable Development Goals (SDGs) are linked together, and their goal is to deal with the root causes of poverty and injustice while also promoting long-term economic growth and protecting the environment. They know that reducing poverty and achieving sustainable development require governments, corporations, civil society, and people to work together.

The SDGs are like a road plan that governments and other groups can use to make sure that their policies, programmes, and investments lead to sustainable development. They want integrated methods that balance economic, social, and environmental factors and take into account the needs of both the present and the future.

For each SDG goal, indicators have been set up so that success can be tracked. Governments, companies, and civil society groups keep track of and report on what they are doing to reach the goals. This is done to be open and accountable for what they are doing. The SDGs have become a way for global and local projects to work together to make the world more fair, inclusive, and sustainable by 2030.

4.4: The Global Reporting Initiative (GRI) Standards

The Global Reporting Initiative (GRI) is a non-profit international group that has made a full structure for reporting on sustainability. The goal of the Global Reporting Initiative (GRI) is to promote sustainable development by giving groups a standardised and globally accepted structure for measuring and reporting their economic, environmental, and social effects.

Since it was started in 1997, GRI has become one of the most widely used and respected sustainability reporting methods in the world. GRI's structure is based on the idea that organisations must be open and accountable about how they do in terms of sustainability so that stakeholders can measure their effects and make smart choices.

The GRI Structure is made up of reporting rules that tell organisations how to report on their performance in terms of the economy, the environment, and society. These guidelines are updated and changed often to keep up with new sustainability problems and the needs of stakeholders.

Reporting Principles: The GRI Structure is based on a set of reporting principles that guide the process of reporting. These concepts include how important the information is, how it affects the stakeholders, how it relates to sustainability, how complete, balanced, comparable, accurate, up-to-date, and clear it is. They make sure that organisations report information that is useful and covers the most important effects and concerns of stakeholders.

Sustainability Reporting Standards: GRI has made a set of Sustainability Reporting Standards that give specific guidelines and indicators for reporting on different sustainability issues. The guidelines cover economic, environmental, and social aspects, and they are put together in sets that are all about the same thing. For example, there are guidelines for reporting on governance, human rights, labour practises, greenhouse gas emissions, water, biodiversity, and a lot of other things.

Levels of Reporting: GRI has three levels of reporting: Core, Comprehensive, and Materiality Disclosures Only. These levels give organisations options based on how far along they are in their reporting and how much information they want to share. At the Core level, organisations have to report on a basic set of indicators. At the Comprehensive level, organisations have to report on a wider range of indicators. Materiality Disclosures Only only reports on important topics that have been found through a strong materiality review.

Materiality Assessment: GRI says that reporting on Materiality is important. Materiality assessment helps organisations figure out which economic, environmental, and social issues are most important to their stakeholders and have the biggest effect on sustainable development. It also helps them put those issues in order of importance. GRI gives advice on how to do materiality reviews so that organisations can make sure they report on the most important things.

Engagement of Stakeholders: The GRI stresses how important it is for stakeholders to be involved in the reporting process. Engaging with stakeholders helps organisations figure out what the most important problems are, collect the right data, and understand the views and worries of stakeholders. GRI gives advice on how to find stakeholders, how to get them involved, and how to talk to them in a good way.

Reporting Process: GRI suggests that organisations use a structured way to report. This includes setting the scope and limits of the report, gathering data and information, analysing and understanding the results, and writing up what was found. GRI's Structure tells organisations what to do at each step of the reporting process so that they report in a uniform and strong way..

By adopting GRI's complete Structure, organisations can make sure their ways of reporting are in line with internationally accepted standards and principles. This makes it possible for stakeholders to look at an organization's performance, growth, and commitment to sustainable development by making sustainability reporting transparent, comparable, and accountable.

The GRI Structure covers a wide range of sustainability topics, such as governance, human rights, labour practises, climate change, biodiversity, product responsibility, community involvement, and more. The rules give different types and sizes of organisations different ways to report, so reporting can be done in a way that fits each organization's needs.

GRI reporting pushes companies to do more than just measure and report their effects. It also talks about how important it is to involve stakeholders, figure out what's important, and set goals for improving performance. By

     The Comprehensive Guide On CSR For Consultant

following the GRI guidelines, organisations can improve their credibility, openness, and sense of responsibility when it comes to problems of sustainability.

GRI reporting gives important information to stakeholders, such as investors, customers, employees, non-governmental organisations (NGOs), and governments, so they can judge an organization's performance in terms of sustainability and make smart choices. It also makes it easier for organisations to compare themselves to each other and use benchmarking, which drives the continuous growth of sustainability practises.

4.5: OECD Guidelines for Multinational Enterprises

The Organisation for Economic Cooperation and Development (OECD) Guidelines for Multinational Enterprises, also called the OECD Guidelines, are a set of suggestions made by the OECD. These tips give multinational companies (MNEs) guidelines and standards that they can follow if they want to do business in an ethical way.

The OECD Guidelines were made in 1976 and have been changed many times since then to keep up with changing business and public standards. The rules apply to all MNEs, no matter what business they are in or how big they are, that operate in or from countries that follow the recommendations.

The goal of the OECD Guidelines is to get MNEs to do business in a responsible way and to give them a way to deal with social, environmental, and governance issues connected to their operations. The guidelines are based on the ideas of internationally known standards like the

     The Comprehensive Guide On CSR For Consultant

Universal Declaration of Human Rights, the ILO Declaration on Fundamental ideas and Rights at Work, and the UN Guiding Principles on Business and Human Rights.

Some important parts of the OECD Guidelines are:

General Policies: The guidelines encourage MNEs to make and use policies that are in line with the guidelines' concepts. These policies should address things like human rights, working conditions, the environment, corruption, consumer protection, and sharing information.

Disclosure and Transparency: MNEs are expected to give accurate and timely information about their activities, structure, financial position, and performance. This includes being open about the effects on society and the environment, how decisions are made, and how stakeholders are involved.

Human Rights: The guidelines emphasise how important it is for multinational enterprises (MNEs) to respect human rights in all of their activities and supply chains. This means not being a party to human rights abuses, backing the rights to freedom of association and collective bargaining, and taking care of the rights of indigenous peoples.

Employment and Industrial Relations: The standards encourage fair employment practises, non-discrimination, and respect for workers' rights. They urge MNEs to give workers safe and healthy places to work, fair wages, and chances to learn and grow.

Environment: MNEs are supposed to use practises that are good for the environment, such as preventing pollution,

 The Comprehensive Guide On CSR For Consultant

using resources efficiently, and taking care of their waste and emissions in a responsible way. The rules also say that businesses should help protect the environment and keep biodiversity alive.

Bribery and corruption: The rules stress how important it is to stop bribery, corruption, and other unethical ways of doing business. MNEs are urged to make policies against corruption, support fair competition, and back efforts to stop corruption.

A network of national contact points (NCPs) is used to carry out the OECD Guidelines. Each country that agrees to follow the standards names an NCP to act as a point of contact for stakeholders who have questions about the rules. NCPs deal with specific claims of not following the rules and help partners and MNEs talk to each other.

By following the OECD Guidelines, MNEs can show they care about doing business in a responsible way and add to sustainable development. The rules set up a structure for meeting community standards and promoting good practises across borders. This makes it easier for MNEs to contribute to economic, social, and environmental well-being in a positive way.

4.6: International Labor Organization (ILO) Conventions

The International Labour Organisation (ILO) is a branch of the United Nations that sets international standards for work and works to support social justice and decent work around the world. The International Labour Organisation (ILO) has made a number of conventions and recommendations that serve as international labour

   The Comprehensive Guide On CSR For Consultant

standards and address a wide range of issues linked to work. These treaties are legally binding agreements that should be ratified and carried out by all member countries.

The ILO conventions cover a wide range of problems related to workers' rights, working conditions, and social protection. Important ILO agreements include:

Freedom of Association and Protection of the Right to Organise (Convention No. 87): This convention recognises that workers and employers have the right to form and join organisations of their own choice, without interference. It also protects the right to deal as a group and sets up protections against anti-union bias.

Right to Organise and Collective Bargaining (Convention No. 98): This convention goes along with Convention No. 87 by giving workers and employers specific rights when it comes to collective bargaining and encouraging good negotiations.

Forced Labour (Conventions No. 29 and 105): Convention No. 29 calls for the end of all kinds of forced or compulsory labour, such as slavery, debt bondage, and human trafficking. Convention No. 105 is about getting rid of forced labour in certain situations.

Minimum Age (Convention No. 138) and Worst Forms of Child Labour (Convention No. 182): Convention No. 138 sets the minimum age for work or employment and the rules for protecting young workers. Convention No. 182 lists the worst forms of child labour. Convention No. 182 says that the worst kinds of child labour, like dangerous work and abuse, should be stopped.

   The Comprehensive Guide On CSR For Consultant

Equal Remuneration (Convention No. 100): This convention works to make sure that men and women get the same pay for the same work, without any gender inequality. It aims to get rid of pay differences between men and women and make sure that everyone gets paid fairly.

Occupational Safety and Health (Convention No. 155): This convention outlines the rules and principles that should be followed to make sure a workplace is safe and healthy. It talks about things like dangers in the workplace, preventing accidents, and making sure that the right safety and health measures are in place.

Decent Work for Domestic Workers (Convention No. 189): This convention aims to improve the working conditions and social security of domestic workers by recognising their rights and promoting equal treatment with other workers.

These are just a few of the many ILO agreements that protect workers' rights and social welfare. The ILO also puts out suggestions that give more help on how to put these conventions into practise and solve problems that are coming up in the workplace.

The ILO conventions have a big effect on the rules and practises of work around the world. They give member countries a way to make and use labour standards that promote good work, social fairness, and worker safety all over the world.

4.7: Carbon Disclosure Project (CDP)

     The Comprehensive Guide On CSR For Consultant

The Carbon Disclosure Project (CDP) is an international non-profit organisation that runs a global disclosure system that lets companies, cities, states, and regions measure, manage, and share their environmental impacts, especially those related to climate change. CDP works with investors, businesses, and governments to support openness and accountability in the face of climate change and other environmental threats.

CDP was started in 2000 to encourage companies to talk about their greenhouse gas emissions and strategies for dealing with climate change. Its focus has grown over time to include water and tree loss, among other environmental problems. The main goal of CDP is to collect and share data and information about the environment to help with making decisions, managing risks, and running a business in a sustainable way.

The most important parts of CDP's work are:

Climate Change: The CDP's climate change programme asks companies and other organisations to share information about their greenhouse gas emissions, energy use, and plans for dealing with climate change. Investors, stakeholders, and policymakers use this information to figure out the risks and possibilities that different organisations face because of climate change.

Water Security: The CDP's water security programme wants to improve the way water is managed and shared by asking for information on how much water is used, what risks are involved, and what strategies are used. This helps businesses and organisations find water-related risks, like

   The Comprehensive Guide On CSR For Consultant

lack of water and pollution, and come up with plans for managing water in a healthy way.

Forests: The CDP's forests programme works on stopping trees from being cut down and keeping forests safe. It encourages companies to talk about their effects on forests, such as commodity-driven deforestation in supply chains, and it backs efforts to promote sustainable buying and production.

CDP works on an annual schedule, and businesses and other groups answer a questionnaire about a wide range of environmental problems. The information is made public through CDP's database and papers, so investors, consumers, and other interested parties can look at how the environment is doing and how it is changing over time.

Investors, financial institutions, governments, and organisations use CDP data and insights to find environmental risks and possibilities, make decisions that take sustainability into account, and improve environmental performance. The CDP disclosure platform gives organisations a standard way to measure and report on their environmental effects. This makes it possible to compare and benchmark.

4.8: United Nations Guiding Principles on Business and Human Rights

The United Nations Guiding Principles on Business and Human Rights (UNGPs) are a set of international rules that explain what states and companies are responsible for when it comes to human rights. The UN Human Rights Council approved the UNGPs in 2011. They are the global structure for dealing with how business actions affect human rights.

     The Comprehensive Guide On CSR For Consultant

The UNGPs rest on three main points:

The State's Duty to Protect Human Rights (It's The Government's Job To Protect People's Rights)	States are in charge of protecting people from third parties, like companies, who break human rights.
	States should make and enforce laws and rules that tell businesses that they have to protect human rights.
	States should give people who have their human rights violated because of business access to good solutions.
The Corporate Responsibility to Respect Human Rights (It's A Company's Job To Respect People's Rights)	Human rights should be respected by all businesses, no matter what size, industry, or location they are in.
	Businesses should do their "due diligence" when it comes to human rights to find, stop, lessen, and report on how they affect human rights.
	Businesses should make policies and processes that show how much they care about human rights and tell people about them.
Access to Remedy (Ability to Get Help)	Both the government and companies should make sure that people who have their rights violated have access to effective legal and non-legal remedies.
	States should make sure people have access to effective legal systems, and businesses should set up ways for people to voice worries about human rights.
	Businesses should work together to fix human rights violations, such as by making amends and making sure

<table>
<tr><td></td><td>people are held accountable.</td></tr>
</table>

These concepts give businesses a complete plan for how to think about human rights when running their businesses. The UNGP stresses how important it is to stop violations of human rights, give people a way to get their rights back, and set up effective ways to deal with problems between business activities and human rights.

4.9: Accountability's AA1000 Series of Standard

The AA1000 series of standards is a set of rules and structures made by AccountAbility, an international group that focuses on sustainability and business responsibility. These standards give organisations a way to measure, manage, and talk about their sustainability success and how they work with stakeholders.

The three main standards in the AA1000 group are:

AA1000 Principles: The AA1000 Principles are the basis of the series and give a set of basic rules for managing and reporting on sustainability. Some of these concepts are openness, relevance, responsiveness, and impact. They help organisations figure out what their most important sustainability problems are, how to deal with them, and how to involve stakeholders.

AA1000 Assurance: The AA1000 Assurance Standard tells you how to do assurance activities on information and processes related to sustainability. It lays out the rules and procedures for independent verification and validation of green performance, making sure that reporting is honest and accurate.

AA1000 Stakeholder Engagement: The AA1000 Stakeholder Engagement Standard offers guidance on effective stakeholder engagement methods. It helps organisations find and talk to stakeholders, learn about their concerns and expectations, and use feedback from stakeholders to make decisions and build strategies.

The AA1000 series encourages organisations to build sustainability into all of their management processes and stresses the importance of responsibility, openness, and interaction with stakeholders. It encourages organisations to look at their sustainability performance as a whole, taking into account social, environmental, and governance issues, and to share this information with stakeholders in an effective way.

The AA1000 series is flexible and can be used in many different ways, so organisations can make their applications fit their wants and situations. It can be used by businesses, non-profits, government agencies, and other groups that want to improve their environmental performance and how they work with stakeholders.

By using the AA1000 series, organisations can strengthen their sense of responsibility, learn more about the effects of sustainability, and build valuable relationships with their stakeholders. The standards help businesses move towards their sustainability goals and promises by building trust and credibility both inside and outside the organisation.

4.10: Social Accountability International (SAI): SA8000 Standard

SA8000 is a standard for social responsibility in the workplace that is known all over the world. It was made by a non-profit group called Social Accountability International (SAI), which works to improve human rights and working conditions in the global supply chain. SA8000 is based on widely recognised labour standards and tries to make sure that workers are treated fairly and ethically.

SA8000 talks about a wide range of social responsibility problems, such as workers' rights, health and safety, working hours, child labour, forced labour, freedom of association, discrimination, and pay. The standard sets up a structure for organisations to put in place policies, processes, and practises that protect workers' rights and promote ethical business practises.

SA8000's most important qualities are:

Child Labour and Forced Labour: SA8000 forbids the use of child labour (below the age of 15 or the legal working age, whichever is higher) and forced labour in any form, including debt bondage and involuntary slavery.

Health and safety: The standard says that companies have to make sure their workplaces are safe and healthy. They have to find and fix any health and safety risks at work, and they have to make sure their employees have access to the safety gear they need.

Freedom of Association and Collective Bargaining: SA8000 protects the rights of workers to freely join or start trade unions and to bargain as a group. It stresses how important it is for companies and workers to talk to each other and work together in a positive way.

Discrimination and Equal Opportunity: Organisations are expected to support equal opportunity, end

   The Comprehensive Guide On CSR For Consultant

discrimination based on race, gender, religion, disability, or other protected characteristics, and make sure that all workers are treated fairly and equally.

Working Hours: SA8000 limits how long you can work, encourages sensible work schedules, and makes sure you get breaks and time off.

Management Systems and Openness: SA8000 needs organisations to set up and keep up an effective social management system, which includes policies, procedures, and ways to keep track of how well the system is working. It also talks about how important openness and involving stakeholders are.

Organisations that want to be SA8000-accredited must let approved certification bodies do independent audits. With these audits, the company is checked to see if it meets the SA8000 standard and if there are enough social accountability measures in place.

SA8000 is a standard for responsible business practises and supply chain management that is widely known and used by businesses, vendors, and other stakeholders. It helps companies show that they care about doing the right thing, taking care of their workers, and being socially responsible.

4.11: OECD CSR Policy Tools

The Organisation for Economic Cooperation and Development (OECD) has made a number of policies to support corporate social responsibility (CSR) in its member countries and beyond. These tools are meant to help governments, businesses, and other stakeholders

    The Comprehensive Guide On CSR For Consultant

incorporate CSR principles and practises into their policies and operations. Here are some of the most important CSR policy tools used by the OECD:

OECD Guidelines for Multinational Enterprises: The OECD Guidelines tell international businesses how to act in a way that is good for business. They cover a wide range of topics, such as human rights, worker rights, the environment, fighting corruption, protecting customer interests, and making information public. The governments of OECD member countries have agreed to support the Guidelines, which are used as a guide for supporting responsible business behaviour around the world.

National Action Plans on Business and Human Rights: The OECD helps countries make and use National Action Plans (NAPs) on Business and Human Rights. These NAPs show how the UN Guiding Principles on Business and Human Rights can be incorporated into state policies and rules. The OECD helps countries make and use NAPs by giving them advice and making it easy for them to learn from each other.

Due Diligence Guidance for Responsible Business Conduct: The OECD has made a number of papers on due diligence that give businesses practical advice on how to act in a responsible way. There is the Due Diligence Guidance for Responsible Supply Chains in the Garment and Footwear Sector, the Due Diligence Guidance for Meaningful Stakeholder Engagement in the Extractive Sector, and the Due Diligence Guidance for Responsible Agricultural Supply Chains.

 The Comprehensive Guide On CSR For Consultant

OECD Tools for Responsible Business Conduct (RBC):
The OECD has made a set of useful tools and materials to help businesses act in a responsible way. Some of the things that these tools cover are supply chain due diligence, risk assessment, stakeholder involvement, and ways to handle complaints. The OECD Risk Awareness Tool for Multinational Enterprises in Weak Governance Zones and the OECD Due Diligence Guidance for Responsible Supply Chains of Minerals from Conflict-Affected and High-Risk Areas are two examples of these tools.

Global Forum on the Responsible Conduct of Business:
The Global Forum is a place where governments, companies, trade unions, and members of civil society can talk about and promote ethical business practises. The platform makes it easier for people to talk about CSR, share what they know, and learn from each other. It gives partners a place to share good business practises, talk about problems, and come up with ways to work together to promote responsible business behaviour.

These OECD CSR policy instruments are very important for helping governments and companies use CSR ideas in their plans, policies, and operations. They want to support ethical business practises, encourage sustainable development, and deal with social and environmental problems that come up because of business.

4.12: The Social Return on Investment (SROI) Structure

SROI is a structure and method for measuring and valuing the social, environmental, and economic results of a group, project, or intervention. In addition to financial indicators,

   The Comprehensive Guide On CSR For Consultant

it is a way to measure the effect and value on a broader scale.

The goal of SROI is to track and measure both the planned and unintended positive and negative results of an organization's actions. It takes into account the opinions and views of stakeholders, especially those who are affected by the work of the organisation. The goal is to understand and describe the social value created, which will give a full picture of how successful an organisation is.

Stakeholder identification and engagement: Identify and engage with stakeholders who are affected by or have an interest in the organization's activities. This ensures that a broad range of perspectives and impacts are considered in the evaluation.

Mapping and valuation of outcomes: Identify the social, environmental, and economic outcomes resulting from the organization's activities. These outcomes are then valued using appropriate methods, such as financial proxies or valuation techniques that reflect stakeholder preferences.

Impact assessment and data collection: Assess the changes or impacts that can be attributed to the organization's activities. This step involves collecting and analyzing data to measure the outcomes and changes experienced by stakeholders.

Monetization and calculation of the SROI ratio: Assign a monetary value to the outcomes identified in order to calculate the SROI ratio. The SROI ratio is expressed as a ratio of social value created compared to the resources invested.

   The Comprehensive Guide On CSR For Consultant

Reporting and communication: Present the findings of the SROI analysis in a clear and transparent manner. This includes sharing the methodology, assumptions, and limitations of the analysis. The SROI report can be used to communicate the organization's social impact to stakeholders and inform decision-making.

SROI provides a complete understanding of social impact, supports evidence-based decision-making, facilitates stakeholder participation, and improves accountability and transparency.

It's vital to remember that SROI has its own set of challenges and limits. It needs subjective judgement in evaluating social consequences, relies on accessible data, and may encounter attribution and causality challenges. SROI, on the other hand, can provide useful insights into the social value provided by organisations and help to better social and environmental outcomes when executed carefully and transparently.

4.13: The London Benchmarking Group (LBG) Model

LBG stands for "London Benchmarking Group," which is a global network of businesses that share best practises and work together on community funding and corporate social responsibility (CSR) reporting. The LBG model was made by the London Benchmarking Group in the early 1990s, and it is now a widely known and used method.

The LBG model gives businesses a standard way to track and share their community investments in a way that is easy to compare. It looks at how much money goes into community projects (like donations) and how much comes

 The Comprehensive Guide On CSR For Consultant

back (like the number of people helped and volunteer hours). The key components of the LBG model include:

Inputs: This refers to the financial investments made by the company in community initiatives. It includes direct cash contributions, in-kind donations, employee volunteering, and management costs associated with community investment activities.

Outputs: Outputs capture the tangible results or activities delivered through community investments. This includes the number of beneficiaries reached, the hours of employee volunteering, the value of in-kind donations, and other relevant metrics.

Outcomes: Outcomes represent the changes or impacts resulting from the community investments. It involves understanding the difference made to the beneficiaries or communities and assessing the positive changes brought about by the company's initiatives.

The LBG model encourages businesses to include community investment in their business plans and goals for making a positive social effect. It gives businesses a structured way to measure the effectiveness and value of their community investments. This makes it easy to make decisions and allocate resources.

Companies can use the LBG model to keep track of and report on their community investments over time. This lets them compare themselves to others in their field and show how far they've come towards their social and community impact goals. The method sets up a consistent structure and language for CSR reporting, which improves transparency, accountability, and stakeholder participation.

It's important to keep in mind that the LBG model has changed over time to keep up with changes in CSR reporting and business community investment. So, if you want the most up-to-date information on the LBG model, you should look at the London Benchmarking Group's most recent suggestions and changes.

4.14: National Voluntary Guidelines on Social Environmental and Economic responsibilities of business

The National Voluntary Guidelines on Social, Environmental, and Economic Responsibilities of Business (NVGs) are a set of guidelines developed by the Ministry of Corporate Affairs, Government of India. These guidelines aim to provide a Structure for businesses to integrate social, environmental, and economic considerations into their core strategies and operations.

The NVGs were first introduced in 2011 and were revised in 2019 to align with international standards and emerging best practices. They are based on the principles of sustainability, inclusivity, and responsible business conduct.

Key features and principles of the National Voluntary Guidelines include:

Ethical Conduct	Businesses should conduct themselves with integrity, transparency, and fairness. They should uphold ethical standards in their operations, including in dealings with stakeholders.
Corporate	Businesses should establish and

    The Comprehensive Guide On CSR For Consultant

Governance	maintain effective systems of corporate governance. They should ensure accountability, fairness, and protection of the rights of shareholders and stakeholders.
Protection of Human Rights	Businesses should respect and promote human rights within their sphere of influence. They should not engage in practices that violate human rights and should address any adverse impacts.
Environmental Sustainability	Businesses should adopt sustainable practices and minimize their environmental footprint. They should conserve resources, manage waste effectively, and contribute to environmental sustainability.
Customer Focus	Businesses should prioritize customer satisfaction and fair business practices. They should provide products and services that are safe, reliable, and meet customer needs.
Reporting and Disclosure	Encourages businesses to disclose relevant social, environmental, and economic information in their annual reports, thereby enhancing transparency and accountability.
Supplier and Partner Relationships:	Businesses should establish fair and transparent relationships with suppliers and partners. They should promote responsible business practices throughout their value chains.

    The Comprehensive Guide On CSR For Consultant

Employee Well-being	Businesses should ensure the well-being, safety, and development of their employees. They should promote equal opportunities, diversity, and non-discrimination.
Social and Community Development	Businesses should contribute to the socio-economic development of the communities in which they operate. They should engage in responsible philanthropy and support community initiatives.
Responsible Business Practice	Businesses should promote responsible innovation, research, and development. They should adhere to principles of fair competition and avoid unfair trade practices.

It is important to note that the NVGs are unique to India and are based on the country's needs and goals. Businesses in India can use these rules to make their business practises more sustainable and moral.

   The Comprehensive Guide On CSR For Consultant

CHAPTER 5: CORPORATE SOCIAL RESPONSIBILITY IN INDIA: A BRIEF HISTORY

In India, the concept of Corporate Social Responsibility (CSR) has evolved over several decades, driven by both societal and legal changes. The following is a brief history of CSR in India:

Pre-Independence Era: Philanthropy and philanthropic efforts by corporate leaders were common during the pre-independence period. Many industrialists and business owners contributed significantly to social causes such as education, healthcare, and community development.

Post-Independence Period (1950s-1990s): After India's independence in 1947, the emphasis turned to nation-building and economic development. The government was central to driving social welfare projects, while companies were primarily concerned with economic growth. During this time, public sector undertakings (PSUs) played an important role in conducting social and developmental activities as part of their commercial operations.

Voluntary Initiatives Emerge (1990s): With the 1990s' economic liberalisation and rise of the private sector, there was a growing realisati

on of the need for firms to go beyond profit and contribute to societal wellbeing. Companies' voluntary initiatives, such as environmental protection, community development, and employee welfare programmes, began to acquire traction.

Corporate Governance and CSR Provisions (2000s): The Companies Act, 1956 mandated the appointment of corporate social responsibility committees for certain classes of companies. However, CSR remained largely voluntary until the introduction of Section 135 in the Companies Act, 2013. This amendment made CSR spending mandatory for companies meeting specific financial thresholds. It required qualifying companies to allocate a percentage of their profits towards CSR activities and established reporting and disclosure requirements.

National Voluntary Guidelines (2011): The National Voluntary Guidelines on Social, Environmental, and Economic Responsibilities of Business were introduced in 2011 by the Ministry of Corporate Affairs, Government of India. These principles gave corporations a Structure for incorporating social and environmental factors into their primary strategies and daily operations. Stakeholder involvement, responsible governance, and environmentally friendly corporate practises were emphasised.

Revised Companies Act and Enhanced CSR Provisions (2014): In 2014, the Companies Act was further amended to enhance CSR provisions. The amendments expanded the scope of CSR activities, provided greater flexibility in choosing projects, and increased reporting requirements. The law also required companies to establish a CSR committee, formulate a CSR policy, and report on CSR activities in their annual reports.

   The Comprehensive Guide On CSR For Consultant

Impact of Global Initiatives and Reporting Structures: Global CSR initiatives, such as the United Nations Global Compact and the Sustainable Development Goals (SDGs), have influenced the CSR landscape in India. Reporting Structures, such as the Global Reporting Initiative (GRI), have provided guidance on CSR reporting and disclosure practices.

CSR is becoming a key component of company strategy and business strategy in India. Companies from a range of industries actively participate in CSR programmes, making contributions to a variety of causes including those including women's empowerment, the environment, healthcare, education, and rural development.

It's crucial to remember that India's CSR development is a continuous process, and new social and environmental issues are being addressed by evolving statutory and voluntary Structures.

   The Comprehensive Guide On CSR For Consultant

CHAPTER 6: SUSTAINABILITY AND CORPORATE SOCIAL RESPONSIBILITY

Sustainability and corporate social responsibility (CSR) are closely related and mutually supportive concepts. Both CSR and sustainability aim to promote ethical business practises and enhance both social and environmental conditions. They share objectives regarding ethical conduct, social advancement, environmental protection, and long-term profitability.

6.1: Triple Bottom Line Structure

CSR and sustainability adopt a holistic approach by considering the triple bottom line: economic, social, and environmental aspects. They recognize that businesses should not focus solely on financial performance but also consider their social and environmental impacts.

The Triple Bottom Line (TBL) is a Structure that measures a company's performance not only in terms of its financial profits but also in terms of its social and environmental impacts. It goes beyond traditional financial reporting and takes into account three interconnected pillars: People, Planet, and Profit. The TBL concept was introduced by John Elkington in 1994, and it aims to encourage businesses to consider their broader responsibilities and contributions to society and the environment.

The three components of the Triple Bottom Line are as follows:

People (Social): This dimension focuses on the social impact of a company's activities on its employees, customers, suppliers, communities, and other stakeholders. It considers labour practises, employee well-being, diversity and inclusion, community engagement, and philanthropy. A socially responsible company seeks to create positive social outcomes and improve the well-being of its stakeholders.

Planet (Environmental): The environment dimension looks at how a company's activities and supply chain affect the environment. It involves measuring the company's carbon footprint, how much energy and resources it uses, how much trash it makes, how much water it uses, and how hard it works to be sustainable and save resources. Businesses that care about the environment try to keep their negative effects on the environment to a minimum and help protect and improve the environment.

Profit (Economic): The economic dimension is the Tripal bottom line, which shows how well a business is doing financially and how much money it is making. It includes income, costs, profits, and other measures of a business's financial health. A business needs to be able to stay in business and make money so it can keep helping people and the world.

The Triple Bottom Line Structure encourages businesses to consider the interconnectedness of social, environmental, and economic factors when making decisions. Companies can create long-term value and contribute to sustainable

   The Comprehensive Guide On CSR For Consultant

development by balancing the interests of all stakeholders and being cognizant of their environmental impacts. TBL is frequently associated with Corporate Social Responsibility (CSR) principles and sustainable business practises.

6.2: Business for Social Responsibility (BSR) Five-Step Approach

Business for Social Responsibility (BSR) is a global non-profit organisation that collaborates with businesses and other stakeholders to promote responsible and sustainable business practises. BSR's mission is to create a more inclusive and sustainable global economy through collaboration with businesses. BSR offers a variety of services, such as research, consulting, and opportunities for collaboration, to assist businesses in integrating sustainability into their fundamental business strategies and operations.

The BSR Structure, also known as the "BSR Five-Step Approach," is a framework that enables businesses to effectively address sustainability challenges and opportunities. It provides a comprehensive and structured approach to incorporating sustainability into business practises. The five BSR Structure stages are as follows:

Assess	Assessing the company's current sustainability performance and identifying sustainability risks and opportunities is the initial step. This involves analysing ESG impacts and comprehending stakeholder expectations.
Define	Based on the findings of the assessment, the company defines its sustainability strategy, objectives, and goals at this

    The Comprehensive Guide On CSR For Consultant

	stage. It involves establishing sustainability objectives that align with the mission and vision of the organisation.
Integrate	This phase focuses on integrating sustainability considerations into the organization's fundamental business processes, decision-making, and value chain. This requires the incorporation of sustainability principles throughout the organisation and the participation of employees at all levels.
Implement	The implementation phase entails taking action to attain the sustainability objectives and targets that have been established. This includes the development of action plans, the allocation of resources, and the implementation of initiatives and projects designed to effect positive change.
Engage	The final step focuses on engaging stakeholders, such as customers, suppliers, communities, and investors, to develop strong relationships and collaborate on sustainability issues. Effective stakeholder engagement is essential for understanding their concerns and expectations and addressing them accordingly.

Throughout these stages, BSR encourages a collaborative approach, involving a variety of stakeholders in order to effect significant and long-lasting change. The Structure is adaptable and malleable across industries, organisational

   The Comprehensive Guide On CSR For Consultant

sizes, and geographies. It is intended to assist businesses in enhancing their sustainability performance, driving innovation, and creating shared value for their business and society.

6.3: The Principles for Responsible Investment (PRI)

The Principles for Responsible Investment (PRI) is a global initiative launched in 2006 by the Principles for Responsible Investment Association, which is supported by the United Nations. The PRI is a network of international investors committed to integrating environmental, social, and governance (ESG) considerations into investment decision-making and ownership practises. The objective of the initiative is to promote responsible investment practises and contribute to a more sustainable global financial system.

The PRI is comprised of six voluntary principles that provide investors with a structure for integrating ESG considerations into their investment processes and engaging with companies to improve their sustainability performance. The six fundamentals are as follows:

Principle 1: We will incorporate ESG issues into investment analysis and decision-making processes.: Signatories commit to considering material ESG factors when evaluating investment opportunities, assessing risk, and making investment decisions.

Principle 2: We will be active owners and incorporate ESG issues into our ownership policies and practices: Signatories commit to engaging with companies they invest in to encourage responsible business practices and transparency on ESG matters.

   The Comprehensive Guide On CSR For Consultant

Principle 3: We will seek appropriate disclosure on ESG issues by the entities in which we invest.: Signatories encourage companies to disclose relevant ESG information to enable informed investment decision-making.

Principle 4: We will promote acceptance and implementation of the Principles within the investment industry: Signatories commit to promoting responsible investment practices and encouraging other investors to adopt the PRI.

Principle 5: We will work together to enhance our effectiveness in implementing the Principles.: Signatories collaborate with other investors and stakeholders to share knowledge, best practices, and lessons learned on responsible investment.

Principle 6: We will each report on our activities and progress towards implementing the Principles.: Signatories are required to publicly report on their responsible investment activities and progress annually.

Institutional investors demonstrate their commitment to responsible investing and support for sustainable development by signing the PRI. The PRI Structure provides a comprehensive and systematic approach to incorporating ESG factors into the investment process. It includes equities, fixed income, infrastructure, and private equity among other asset classes.

The PRI initiative has grown substantially over the years, and as of my most recent update in September 2021, it has thousands of signatories representing trillions of dollars in managed assets. The initiative continues to play an essential role in advancing responsible investment

   The Comprehensive Guide On CSR For Consultant

practises, engaging with companies on sustainability issues, and nurturing investor collaboration to create a more sustainable and resilient global financial system.

6.4: The World Business Council for Sustainable Development (WBCSD) Vision 2050

The World Business Council for Sustainable Development (WBCSD) Vision 2050 is a comprehensive and ambitious Structure that outlines a pathway for businesses to achieve a sustainable and equitable world by the year 2050. The Vision 2050 was launched in 2010 by the WBCSD, a global CEO-led organization comprising over 200 leading businesses united in their commitment to sustainable development.

The Vision 2050 Structure is based on the recognition that businesses play a vital role in addressing global challenges, such as climate change, resource depletion, poverty, and social inequality. It sets out a long-term vision for how companies can contribute to a sustainable future by transforming their strategics, operations, and engagement with stakeholders.

Key elements of the WBCSD Vision 2050 include:

Nine Pathways to Sustainability	The Vision 2050 identifies nine pathways that collectively lead to a sustainable world by 2050. These pathways address critical areas such **as energy, water, food, ecosystems, health, education, sustainable cities, and social development.**
Triple Bottom Line Approach	The Structure aligns with the triple bottom line concept, emphasizing the

	importance of economic prosperity, social equity, and environmental stewardship.
Systems Thinking	The Vision 2050 adopts a systems thinking approach, recognizing that complex challenges require holistic solutions that consider the interconnectedness of environmental, social, and economic systems.
Collaboration and Innovation	The Structure underscores the need for collaboration and innovation across sectors and stakeholders to drive sustainable development and overcome global challenges.
Backcasting Approach	The Vision 2050 uses a "backcasting" approach, which involves envisioning a desirable future and then working backward to determine the actions and strategies needed to achieve that future.
Engagement of Business Leaders	As a CEO-led initiative, the Vision 2050 emphasizes the importance of top leadership commitment to drive sustainability initiatives and embed sustainability principles in business strategies.

Vision 2050 provides a framework for businesses to align their operations with sustainable development objectives and contribute to a more inclusive and sustainable global economy. It encourages businesses to develop innovative solutions and collaborate with other stakeholders to accomplish shared sustainability goals.

 The Comprehensive Guide On CSR For Consultant

Over time, the WBCSD has collaborated with its member organisations and other stakeholders to advance the Vision 2050 agenda. The WBCSD seeks to accelerate the transition to a sustainable world by 2050 through various initiatives, projects, and advocacy efforts, while addressing emergent challenges and opportunities along the way.

6.5: The Natural Step

The Natural Step was founded in Sweden in 1989 as a non-profit sustainability structure based on scientific research. It guides organisations in comprehending and addressing complex environmental and social challenges, thereby facilitating their transition to sustainability. The Structure is founded on sustainability science and systems thinking-derived principles.

The Natural Step Structure is founded on four sustainability principles that serve as a compass for decisions and actions:

System Conditions: The Structure outlines four system conditions that, if followed to, would assure a sustainable society. Based on scientific findings, these conditions represent the ecological and social limits that must not be exceeded. These are the system conditions:

- **Substances extracted from the Earth's crust:** The use of non-renewable resources must be minimized to ensure their availability for future generations.
- **Substances produced by society:** Human-made substances should not exceed the Earth's natural capacity to break them down.
- **Degradation of nature:** Human activities should not lead to a reduction in biodiversity or the

     The Comprehensive Guide On CSR For Consultant

capacity of ecosystems to function and renew themselves.

- **Human needs:** All people should have access to resources and opportunities to meet their basic needs, such as clean water, food, and healthcare.

Backcasting: The Natural Step encourages a "backcasting" approach, where organizations envision a sustainable future based on the system conditions and then work backward to identify the steps and strategies needed to achieve that vision.

Principles for Decision-making: The Structure provides a set of principles to guide decision-making and actions towards sustainability. These principles include using a long-term perspective, prioritizing preventive measures, and considering the full life cycle of products and services.

Incremental and Transformative Change: The Natural Step recognizes the importance of both incremental improvements and transformative changes to achieve sustainability goals. While incremental changes address immediate issues, transformative changes lead to fundamental shifts in systems and practices.

The Natural Step Structure has been extensively adopted by businesses, municipalities, educational institutions, and non-profits across the globe. It provides a common language and a structured approach to sustainability, enabling diverse stakeholders to collaborate towards shared objectives.

The Natural Step assists organisations in applying the Structure and incorporating sustainable practises into their operations and decision-making through seminars, training,

 The Comprehensive Guide On CSR For Consultant

consulting, and a variety of sustainability programmes. By adhering to the principles and system conditions, organisations can create value for their stakeholders and the planet while contributing to a more sustainable future.

Both CSR and sustainability require engagement with stakeholders. They acknowledge that businesses must comprehend the requirements, concerns, and expectations of their stakeholders, including employees, customers, communities, non-governmental organisations (NGOs), and governments. CSR promotes responsible conduct consistent with sustainability standards. By incorporating sustainability into their operations, businesses can reduce their environmental impact, conserve resources, promote renewable energy, and implement sustainable supply chain practises.CSR and sustainability both encourage businesses to consider the future. They stress the importance of comprehending the effects of current activities on future generations and maintaining the sustainability of economic, social, and environmental systems.

CSR and sustainability facilitate the identification and management of ESG-related (environmental, social, and governance) risks. By proactively addressing these risks, businesses can strengthen their resiliency, safeguard their reputation, and reduce their potential financial and legal liabilities. The adoption of CSR and sustainability practises can create value for enterprises. By considering social and environmental factors, businesses are able to identify new business opportunities, improve operational efficiency, attract customers and investors, and establish long-term relationships with stakeholders. Both CSR and sustainability stress the significance of reporting and openness. Companies are encouraged to inform stakeholders about their CSR and sustainability initiatives, impacts, and progress. Reporting Structures, such as the Global Reporting Initiative (GRI), offer guidelines for the complete disclosure of CSR and sustainability-related information.

CSR and sustainability are interdependent and mutually supportive. CSR practises promote long-term development by addressing social and environmental concerns, whereas sustainability principles encourage businesses to adopt responsible practises that ensure their long-term viability. By incorporating CSR and sustainability, businesses can produce positive outcomes, manage risks, build trust with stakeholders, and contribute to a more sustainable and inclusive world.

 The Comprehensive Guide On CSR For Consultant

CHAPTER 7: CORPORATE SOCIAL RESPONSIBILITY AND SUSTAINABILITY PROGRAMME IMPLEMENTATION BENEFITS

CSR programmes, also known as Corporate Social Responsibility programmes, are initiatives and activities undertaken by businesses to fulfil their social, environmental, and ethical obligations. These programmes seek to create positive impacts on society, the environment, and stakeholders in addition to maximising profits. CSR programmes can encompass a vast array of topics and activities. Strong Corporate Social Responsibility (CSR) programmes can provide numerous benefits to businesses. Here are several of the most important benefits of instituting robust CSR programmes:

7.1: Improved Reputation and Brand Image

CSR initiatives demonstrate a company's commitment to responsible and ethical business practises. Companies demonstrate their dedication to having a positive impact on society by addressing social and environmental issues. Customers, investors, employees, and other stakeholders gain confidence, and the organization's reputation improves. In today's competitive business world, companies must distinguish themselves from the competition. Companies with effective CSR programmes distinguish themselves by demonstrating their values, mission, and commitment to making a difference. Customers who share these values and prefer to patronise socially responsible businesses may be attracted.

Positive media coverage of CSR initiatives is widespread, both locally and internationally. Media agencies are more likely to cover companies that are actively engaged in social and environmental activities, providing exposure and enhancing the public's perception of the business. Positive media coverage improves the brand's reputation and perception.Customers who are satisfied with a company's CSR initiatives become brand ambassadors, disseminating positive word of mouth and promoting the business. This can result in increased consumer loyalty, a larger customer base, and an improved company reputation.CSR programmes demonstrate a company's commitment to addressing the expectations and concerns of its stakeholders. By interacting with stakeholders and addressing their social and environmental concerns, businesses create stronger relationships and cultivate a positive perception among them, resulting in an improved reputation and brand.

Here are three real-world examples of how instituting CSR and Sustainability programmes has enhanced the reputation and brand image of businesses:

Example

Unilever:

Unilever, a significant consumer goods company, has incorporated sustainability into its business strategy. They have implemented several CSR and sustainability initiatives that have enhanced the company's reputation and brand image.

The Sustainable Living Plan of Unilever, for instance, concentrates on reducing environmental impact, enhancing health and well-being, and enhancing livelihoods. This commitment to sustainable practises has earned Unilever recognition as a leader in sustainability and enhanced the company's reputation as a responsible enterprise.

Dove and Ben & Jerry's, among other Unilever brands, have gained popularity among consumers who value ethical and sustainable products. Their commitment to fair trade, transparency in ingredient sourcing, and initiatives for social and environmental causes have improved their brand image and consumer loyalty..

LEGO:

The Danish toy manufacturer's CSR and sustainability initiatives have strengthened its reputation and brand image. As an illustration, LEGO has ambitious goals for the use of sustainable materials in its products and packaging. Their dedication to reducing plastic waste and promoting environmental stewardship has resonated with consumers, resulting in a favourable perception of the brand as environmentally conscious and accountable.

LEGO's programmes emphasising child development, education, and creativity, such as the LEGO Foundation, have contributed to the company's reputation as one that prioritises the well-being and growth of children. This has

   The Comprehensive Guide On CSR For Consultant

bolstered their reputation as a reputable and socially responsible company.

Microsoft:

Microsoft, the gigantic technology company, has incorporated CSR and sustainability into its corporate strategy, resulting in positive effects on its reputation and brand image. Here are some instances:

Microsoft has made substantial investments in renewable energy with the goal of becoming carbon negative by 2030. Their commitment to sustainability and combating climate change has propelled them to the forefront of the technology industry, enhancing their reputation as an environmentally responsible business.

Microsoft's initiatives to reconcile the digital divide and promote digital inclusion, such as their affordable access programmes and skill development initiatives, have been lauded and recognised. These efforts have enhanced the company's reputation as a socially conscious and technologically empowering enterprise.

In conclusion, companies such as Unilever, LEGO, and Microsoft have demonstrated that implementing effective CSR and Sustainability programmes can contribute to a strengthened reputation and brand image. By aligning their business strategies with sustainability objectives, promoting ethical practises, engaging stakeholders, and addressing social and environmental issues, these businesses have earned the trust and loyalty of consumers, investors, and other stakeholders.

7.2: Increased Stakeholder Engagement

 The Comprehensive Guide On CSR For Consultant

Companies can utilise CSR programmes to address and respond to their stakeholders' concerns. By actively participating in programmes that address social and environmental challenges, businesses demonstrate their commitment to identifying and addressing stakeholder demands and concerns. CSR programmes demonstrate a company's dedication to responsible and sustainable practises, thereby fostering stakeholders' confidence and credibility. Stakeholders perceive companies that actively partake in CSR programmes as more trustworthy and credible, resulting in greater involvement and support.Effective CSR programmes involve stakeholders in decision-making and solicit their input. This can be accomplished through stakeholder consultations, partnerships, advisory groups, and other methods that enable stakeholders to actively participate and contribute to the company's CSR initiatives.

The implementation of CSR programmes generates a platform for open and transparent communication with stakeholders. Companies can provide information about their CSR initiatives, successes, and issues, fostering meaningful dialogue and enhancing stakeholder relationships.CSR programmes frequently involve collaborations and partnerships with stakeholders such as NGOs, community organisations, and government agencies. Companies and stakeholders can collaborate more effectively to address social and environmental issues by leveraging their respective talents, resources, and networks, resulting in increased participation and the creation of shared value.CSR programmes frequently involve collaborations and partnerships with stakeholders such as NGOs, community organisations, and government

     The Comprehensive Guide On CSR For Consultant

agencies. Companies and stakeholders can collaborate more effectively to address social and environmental issues by leveraging their respective talents, resources, and networks, resulting in increased participation and the creation of shared value.

Example

Patagonia, An outdoor clothing and gear brand known for its commitment to environmental sustainability and social responsibility is one real-world example of how CSR programmes have helped increase stakeholder engagement. Several CSR initiatives introduced by Patagonia have increased stakeholder participation. He Here's an instance:

Environmental Measures:

Patagonia places a significant emphasis on environmental sustainability and has implemented numerous initiatives to reduce their ecological imprint. Their "Worn Wear" programme is well-known for promoting product repair, reuse, and recycling. Patagonia encourages customers to repair worn-out apparel and offers a marketplace for purchasing and selling pre-owned Patagonia products. This initiative not only promotes sustainability, but also engages consumers as active participants in the circular economy, thereby strengthening their relationship with the brand and instilling a sense of shared responsibility for environmental stewardship.

Patagonia has taken measures to ensure that its supply chain is transparent and ethical on a global scale. They work closely with their suppliers to promote ethical labour practises, secure working conditions, and environmental responsibility. The "Footprint Chronicles" by Patagonia

 The Comprehensive Guide On CSR For Consultant

provide detailed information about the manufacturing process and social and environmental impacts of their products. By providing this information, Patagonia actively engages stakeholders, including customers and advocacy groups, in understanding and supporting their responsible sourcing practises.

Advocacy & Activism: Patagonia extends its engagement with constituents beyond the scope of its own operations. They actively support environmental advocacy and participate in activism campaigns on issues such as climate change and public lands preservation. By taking a stance on significant social and environmental issues, Patagonia attracts like-minded stakeholders and establishes a community of advocates. This engagement extends beyond consumers to include non-governmental organisations (NGOs), activists, and individuals who share their beliefs, thereby amplifying their effect and influence.

Through these methods, Patagonia has effectively expanded stakeholder participation. Patagonia has established a robust network of stakeholders who are devoted to their mission and values by involving consumers in sustainable practises, promoting supply chain transparency, connecting with local communities, and advocating for crucial issues. This participation not only enhances the company's reputation and brand loyalty, but also promotes positive change and contributes to its long-term success.

7.3: Improved Employee Engagement and Retention

CSR initiatives give workers a sense of purpose and significance in their employment. When an organisation is committed to having a positive impact on society and the environment, employees feel a stronger connection to their work and are more invested in their roles. Frequently, employee values and attitudes are aligned with CSR programmes. When personal values align with those of the organisation, employees are more likely to be engaged and committed to their work. Participation in a socially responsible organisation improves employee morale. Because they enjoy working for a company that supports social and environmental causes, employees are motivated and dedicated to their duties.

Positive workplace cultures that prioritise social responsibility are fostered by CSR programmes. Employees who work in such an environment are more likely to be content with their employment, resulting in greater engagement and retention. CSR programmes frequently include employee participation and involvement. Employees can contribute their skills, time, and ideas to CSR activities, thereby increasing employee engagement. Collaboration and collaboration are frequently necessary for CSR programmes, both internally and with external stakeholders. Participation in CSR activities encourages employees to work together, fostering a sense of teamwork and camaraderie.

As part of CSR initiatives, employees frequently participate in skill development and learning opportunities. By participating in CSR initiatives, employees can acquire new skills, gain diverse experiences, and increase their knowledge, thereby enhancing their professional development and motivation. A robust CSR programme fosters a positive organisational culture. Companies that

    The Comprehensive Guide On CSR For Consultant

prioritise social and environmental responsibility cultivate a culture that is caring, fair, and inclusive, resulting in greater employee engagement and retention. When organisations demonstrate a commitment to CSR, employees feel valued and appreciated. They perceive their company to be a responsible and ethical organisation, which strengthens the employee-employer bond and increases employee retention.

Example

Tata Consultancy Services (TCS), A multinational IT services and consulting firm in India has implemented CSR initiatives to increase employee engagement and retention. TCS has established a variety of CSR initiatives that prioritise employee wellness, skill development, and community engagement. Here's an Instance:

Volunteer Employee Programmes: "TCS Cares" is a significant employee volunteer programme that encourages workers to engage in social activities. Through this programme, employees have the opportunity to donate their time and expertise to various community development initiatives. TCS provides employees with paid time off to volunteer and organises team-based volunteering events. This programme not only encourages employees to donate to charitable organisations, but also instills in them a sense of pride and involvement.

Employee Skill Development and Education: TCS places a premium on the skill development and education of its employees. Several programmes are in existence to assist employees in enhancing their technical and professional

   The Comprehensive Guide On CSR For Consultant

skills. TCS also promotes digital literacy and education in underprivileged communities via programmes such as "TCS IT Wiz" and "TCS iON Digital Learning Hub." TCS promotes a culture of continuous learning and growth by investing in employee skill development and supporting educational initiatives, thereby boosting employee engagement and job satisfaction.

TCS recognises and celebrates its employees for their efforts and accomplishments. There are a number of award programmes in existence, such as the "TCS Long Service Awards" and the "TCS Bravo Awards." These programmes acknowledge employees' commitment and performance, thereby boosting morale and engagement. Additionally, TCS provides opportunities for career advancement and growth, which encourages employees to remain with the organisation and contribute to its long-term success.

These strategies have effectively increased employee engagement and retention at TCS. TCS fosters a pleasant workplace that values and supports its employees by providing volunteer opportunities, prioritising skill development and education, promoting employee well-being and work-life balance, and instituting employee appreciation programmes. This positive culture, fostered by CSR initiatives, increases employee engagement, happiness, and loyalty, thereby contributing to the overall success of the business.

7.4: Risk Mitigation

Strong CSR programmes help in the identification and management of environmental, social, and governance (ESG) risks. By proactively addressing these concerns,

  The Comprehensive Guide On CSR For Consultant

businesses can mitigate potential financial, operational, legal, and reputational risks. This approach to risk management increases the resilience and sustainability of an organisation.

Reputation Risk	CSR programmes help build a good name and image for a brand. When a company does things in a responsible and sustainable way, it is seen as dependable and trustworthy. This helps reduce image risks and the bad effects of any reputation-related crises that might happen.
Regulatory and Legal Risks	CSR programmes make sure that social and environmental laws, regulations, and business standards are followed. By addressing these problems ahead of time, companies reduce the chance of not following the rules, getting fined, or getting into a legal dispute.
Environmental Risks	CSR programmes focus on protecting the earth and using resources in a responsible way. By taking steps to reduce their effect on the environment, businesses reduce the risks associated with climate change, pollution, the depletion of natural resources, and legal requirements related to protecting the environment.
Social and Human Rights	Social issues like labour rights, human rights, diversity and inclusion, and

   The Comprehensive Guide On CSR For Consultant

Risks	community growth are addressed by CSR programmes. Companies reduce the risk of labour disputes, human rights violations, and bad relationships with the community by using responsible practises and treating workers and stakeholders in a fair and honest way.
Supply Chain Risks	CSR programmes encourage responsible buying and the management of the supply chain. Companies can find and deal with possible risks like child labour, forced labour, environmental violations, and supply chain disruptions by evaluating their suppliers' ethical practises, environmental effect, and labour standards.
Financial Risks	CSR programmes help businesses be financially stable and last for a long time. By including environmental, social, and governance (ESG) factors in risk assessment and decision-making processes, businesses can find potential financial risks and deal with them in advance, ensuring long-term profitability and resilience.
Stakeholder Risks	Engaging stakeholders in CSR programmes helps find out what their concerns and demands are and how to meet them. Companies can decrease the risk of stakeholder activism, boycotts, and bad public opinion by understanding the needs of their stakeholders and keeping good

	relationships with them.
Innovation and Adaptation Risks	CSR programs encourage innovation and adaptation to changing societal and environmental demands. By staying ahead of emerging trends and proactively addressing issues, companies reduce the risk of becoming obsolete or losing market relevance.
Supply Chain Transparency	CSR programs promote supply chain transparency, allowing companies to identify and mitigate risks associated with unethical practices, human rights abuses, and environmental violations within their supply chain.
Crisis Management	Companies that implement robust CSR programmes are better prepared for effective crisis management. Companies are better equipped to respond to crises, communicate openly, and mitigate the potential harm to their reputation and operations if they have established responsible practises and strong stakeholder relationships.

Example

Hindustan Unilever Limited (HUL), A well-known Indian consumer products company is an example of how CSR programmes have aided in risk avoidance. HUL has introduced a number of CSR initiatives to address social and environmental issues, thereby reducing risk. Here's an instance:

 The Comprehensive Guide On CSR For Consultant

Management of Sustainable Sourcing and Supply Chain: HUL has developed sustainable procurement practises and responsible supply chain management to mitigate risks associated with environmental and social repercussions. They work closely with their suppliers to attain ethical standards, the responsible procurement of raw materials, and environmental protection. HUL reduces the likelihood of negative environmental impacts such as deforestation and water contamination by promoting sustainable practises throughout their supply chain, as well as addressing social issues such as child labour and unfair working conditions.

Water Conservation and Management: Water scarcity is a major concern for HUL's operations as a manufacturer of consumer goods. HUL has implemented numerous water conservation and management strategies to mitigate this threat. Their "Water for Public Good" initiative, for instance, focuses on restoring water sources and encouraging communities to consume water responsibly. In addition, HUL has made significant strides in reducing its water consumption during production. By managing water scarcity and promoting responsible water management, HUL eliminates operational risks and ensures the future availability of water resources.

Community Development and Engagement: Through their CSR programmes, HUL actively engages with local communities, concentrating on social and economic development. By addressing community requirements and fostering inclusive growth, HUL reduces the likelihood of social unrest and community discontent. Their initiatives include skill development and employment opportunities, health and sanitation promotion, and support for

   The Comprehensive Guide On CSR For Consultant

educational programmes. Through these initiatives, HUL cultivates positive relationships with local communities and fosters an environment conducive to its business operations.

HUL has devised responsible waste management practises to mitigate the environmental risks associated with trash production. They care about reducing waste, promoting recycling and reuse, and ensuring the secure disposal of hazardous materials. The initiatives of HUL, such as their "Clean India" campaign, seek to address waste management issues while promoting community cleanliness and hygiene. HUL reduces the risk of environmental degradation, regulatory noncompliance, and reputational damage by instituting effective waste management procedures.

Through these measures, HUL has successfully addressed risks associated with environmental repercussions, water scarcity, community discontent, and waste management. HUL proactively addresses social and environmental concerns by integrating CSR into their business strategy, thereby mitigating potential risks to their operations and reputation. Not only does this method reduce risk, but it also demonstrates HUL's commitment to sustainable practises, which contributes to the company's long-term viability.

7.5: Access New Markets and Customers

CSR programmes can facilitate entry into new markets and client categories. Many consumers are becoming more conscious of the social and environmental impacts of their purchases. By exhibiting responsible practises and providing sustainable products or services, businesses can attract environmentally and socially conscientious consumers and capitalise on new market opportunities.

CSR initiatives can provide valuable insight into market demand and future trends. Companies can align their products, services, and marketing strategies with the evolving preferences and expectations of potential new markets by recognising the significance of customers' and communities' social and environmental concerns. CSR programmes can help a company separate out from the competition in emerging markets. When businesses actively partake in CSR programmes and communicate their commitment to social and environmental responsibility, they attract customers who value ethical and sustainable business practises. This can give companies a competitive advantage and attract new customers who share their values.

CSR programme implementation frequently requires an understanding of the social and cultural aspects of target markets. By engaging with local communities, identifying their needs, and resolving social challenges unique to that market, businesses gain a better understanding of the cultural environment and develop relationships with prospective consumers. Participating in CSR programmes are customers, community organisations, non-governmental organisations (NGOs), and government authorities. Through these interactions, businesses can build relationships, receive feedback, and gain insights into the preferences and expectations of new markets and consumers. This facilitates the adaptation of products,

    The Comprehensive Guide On CSR For Consultant

services, and communication strategies to market-specific requirements.

Implementing CSR programmes enhances the reputation of a brand and increases consumer trust. Consumers in emerging markets are more likely to have faith in companies with a reputation for social and environmental responsibility. This trust facilitates customer acquisition and retention, particularly in regions where the brand is relatively new. CSR programmes frequently involve collaborations and partnerships with local organisations and constituents in new markets. These partnerships can provide organisations with access to networks, distribution channels, and market data, allowing them to establish a local presence and better understand the requirements and preferences of local consumers.

Participation in CSR initiatives results in devoted and loyal consumers. Customers who concur with a company's principles and appreciate its social and environmental initiatives become brand ambassadors, spreading positive word-of-mouth and attracting new customers in new regions.

Example

ITC Limited

CSR projects have helped ITC Limited, a diverse group with companies in FMCG, hotels, agriculture, and paperboards, reach new markets and customers. ITC has set up a number of CSR projects that focus on sustainable development and giving people more power in their communities. This has helped the company reach new markets and customers. Here is a good example:

 The Comprehensive Guide On CSR For Consultant

Sustainable Agriculture projects: ITC has worked on a number of sustainable agriculture projects that try to make farmers' lives better, spread sustainable farming methods, and improve the quality of agricultural output. Their "e-Choupal" programme gives farmers access to technology and knowledge about the market, the weather, and how to take care of their crops. ITC has earned the trust and loyalty of people in rural areas by helping farmers, pushing sustainable agriculture, and helping farmers in rural areas find new markets and customers.

Community Development Programmes: ITC's corporate social responsibility (CSR) programmes also focus on improving the communities where their businesses are based. They have started programmes to help people in rural and remote places get better health care, education, and sanitation. For example, its "Mission Sunehra Kal" effort tries to make rural places into thriving, self-sufficient communities through a wide range of development projects. ITC has been able to get more customers in these places by meeting the wants of the people who live there and making their lives better.

Products and packaging that are good for the environment: ITC cares about the environment and has made a number of products and packaging solutions that are good for the environment. They are dedicated to reducing their carbon footprint and encouraging others to do the same. For example, ITC's "Paperboards and Specialty Papers Division" offers eco-friendly packaging options to meet the growing market demand for sustainable packaging. ITC has gotten into new market areas by giving clients who care

   The Comprehensive Guide On CSR For Consultant

about sustainability products and services that are good for the environment.

ITC has built a reputation for doing business in an honest way, being open, and having good company governance. Their CSR programmes are in line with these ideas and put a focus on being responsible to society and the environment. Customers who care about ethical brands believe and like ITC because it is committed to doing business in a responsible way. This has helped them reach a wider range of customers and look into new markets where ethical factors are important.

As a result of its CSR efforts, ITC has been able to enter new markets and win new clients. ITC has set itself up as a socially responsible company by focusing on sustainable agriculture, community development, products that are good for the environment, and ethical business practises. This has helped it draw customers who share these values. This has not only helped them get more customers, but it has also made their business more well-known and made them more competitive in the market.

7.6: Innovation and Competitive Advantage

CSR programmes encourage innovation and originality within an organisation. By considering social and environmental factors, businesses are able to identify new opportunities, create sustainable products and services, and adapt to fluctuating market demands. This approach fueled by innovation can provide a competitive advantage in the market.

CSR programmes involve customers, employees, communities, and non-governmental organisations (NGOs).

 The Comprehensive Guide On CSR For Consultant

Through these interactions, businesses gain significant insight into social requirements, expectations, and emerging trends. This stakeholder engagement can assist businesses in identifying new prospects, products, and services that meet societal demands. Businesses that incorporate CSR into their business strategies may distinguish themselves from the competition. By providing new solutions that address social and environmental concerns, businesses can differentiate themselves on the market and attract clients who value ethical and sustainable practises. This distinction provides a competitive edge and establishes the company as the leader in its industry.

CSR programmes encourage businesses to offer eco-friendly goods and services. Examining the entire product lifecycle, from primary material source to disposal, is required. Sustainable product innovators can capitalise on rising consumer demand for environmentally responsible alternatives, giving them a competitive advantage over businesses that do not prioritise sustainability. Frequently, CSR initiatives emphasise resource efficiency and waste reduction. By implementing practises and technologies that minimise environmental impact and maximise resource utilisation, businesses can realise cost savings and operational efficiencies. This could result in a competitive advantage through the delivery of lower-priced or higher-quality goods and services.

CSR programmes enable businesses to anticipate future challenges and respond to emerging trends. By anticipating evolving social, environmental, and regulatory expectations, businesses can innovate proactively and establish themselves as leaders in addressing emergent

 The Comprehensive Guide On CSR For Consultant

issues. This future-proofing enables businesses to maintain a competitive edge and remain relevant in ever-changing business environments.

Example :

Tesla: a The foremost manufacturer of electric vehicles (EV) has revolutionised the automotive industry with its innovative and sustainable approach. Their CSR efforts have significantly contributed to their success:

Innovation in Electric Vehicles: Tesla's CSR initiatives are centred on the development of sustainable transportation solutions to combat climate change. They have pioneered the mass production of electric vehicles and advanced battery technologies, thereby making EVs more accessible and marketable. Their commitment to sustainable transportation and innovation have provided them with a substantial competitive advantage in the automotive industry.

Tesla's involvement in renewable energy extends beyond the production of electric vehicles. They have developed energy storage and solar energy products, such as solar panels and the Powerwall. Tesla has positioned itself as a pioneer in the transition to a sustainable energy future by providing a comprehensive suite of renewable energy solutions.

Tesla's dedication to sustainability and innovation has created a distinctive brand identity and positioned them as a symbol of technological progress and environmental responsibility. Their CSR programmes have increased brand loyalty and attracted environmentally conscious

 The Comprehensive Guide On CSR For Consultant

consumers willing to invest in sustainable transportation alternatives.

Example

Patagonia, A company that manufactures outdoor apparel and equipment is renowned for its commitment to environmental sustainability and social responsibility. Several CSR programmes have contributed to their innovation and competitive advantage:

Patagonia has prioritised the use of sustainable materials and responsible manufacturing processes in its supply chain. They advocate fair trade, organic cotton farming, and the use of recycled materials in their products, thereby reducing their environmental footprint and attracting customers with a concern for the environment.

Worn Wear Programme: Patagonia's Worn Wear programme encourages customers to repair and reuse their clothing rather than purchase new products. This programme not only increases the lifespan of their products, but also promotes a sustainable culture and reduces waste.

Transparency and Advocacy: Patagonia is transparent regarding their environmental and social initiatives, and they engage in environmental advocacy campaigns. Their commitment to openness and advocacy for sustainability have earned them a loyal customer base and a significant competitive advantage.

These examples demonstrate how CSR programmes that emphasise sustainability, innovation, and responsible practises can result in a competitive advantage by appealing

     The Comprehensive Guide On CSR For Consultant

to environmentally conscious consumers, fostering brand differentiation, and driving industry transformation.

7.7 Positive Social Impact

Strong CSR programmes have a positive social impact by addressing social issues, fostering community growth, and empowering the disadvantaged. By investing in education, healthcare, poverty reduction, and other social initiatives, businesses contribute to the well-being and long-term growth of society.

CSR initiatives address social issues such as poverty, inequality, education, healthcare, and community development. By actively participating in programmes addressing these issues, businesses can contribute to positive social change and advance the well-being of individuals and communities. Common CSR programme components include philanthropic initiatives and community investment. Local initiatives, charitable organisations, and community development projects receive resources, finances, and volunteer hours from businesses. By addressing specific requirements and enhancing the quality of life for disadvantaged or marginalised individuals, direct involvement has a positive social impact.

Employment creation and economic growth can be aided by CSR programmes, particularly in communities where businesses operate. Through investing in local economies, promoting entrepreneurship, and providing training and employment opportunities, businesses can foster economic growth and empower individuals and communities. Environmental sustainability CSR programmes have a

  The Comprehensive Guide On CSR For Consultant

positive impact on society. Businesses contribute to mitigating climate change and preserving ecosystems by instituting sustainable practises, reducing carbon emissions, conserving natural resources, and promoting renewable energy, which benefits society as a whole.CSR programmes are beneficial for employees, suppliers, and local communities. By providing equitable working conditions, promoting diversity and inclusion, respecting human rights, and encouraging skill development, businesses enhance the well-being and emancipation of their stakeholders.

Companies contribute to social progress by establishing a culture of honesty, transparency, and respect for human rights through ensuring equitable labour practises, responsible sourcing, and ethical business conduct. When implementing CSR programmes, stakeholders' participation and dialogue are encouraged. By actively involving stakeholders in decision-making processes, businesses ensure that their perspectives are considered and their needs are met. This inclusive strategy yields superior social outcomes and more adaptable solutions to societal issues.

Example:

Coca-Cola:

Coca-Cola, a prominent beverage manufacturer, has implemented CSR programmes with positive social impact:

Coca-Cola prioritises water conservation and replenishment in its water stewardship efforts. Their "Water for Sustainable Development" campaign aims to increase worldwide community access to clean water, sanitation, and hygiene. Together with local individuals and

   The Comprehensive Guide On CSR For Consultant

organisations, they seek water replenishment, water conservation, and watershed protection.

The "5by20" initiative of Coca-Cola promotes women's economic empowerment by providing access to business skills training, financial services, and mentoring. The programme aims to economically empower 5 million women entrepreneurs by the year 2020, thereby fostering gender equality and eradicating poverty.

Coca-Cola participates in numerous community development programmes, with an emphasis on education, health, and communal infrastructure. They provide funding for initiatives that improve education, expand access to healthcare, and foster long-term community development.

Coca-Cola participates in numerous community development initiatives, with an emphasis on education, health, and community infrastructure. They provide funding for initiatives that improve education, expand access to healthcare, and foster long-term community development.

Coca-Cola, a prominent beverage manufacturer, has implemented CSR programmes with positive social impact:

Coca-Cola prioritises water conservation and replenishment in its water stewardship efforts. Their "Water for Sustainable Development" campaign aims to increase worldwide community access to clean water, sanitation, and hygiene. Together with local individuals and organisations, they seek water replenishment, water conservation, and watershed protection.

  The Comprehensive Guide On CSR For Consultant

The "5by20" initiative of Coca-Cola promotes women's economic empowerment by providing access to business skills training, financial services, and mentoring. The programme aims to economically empower 5 million women entrepreneurs by the year 2020, thereby fostering gender equality and eradicating poverty.

Coca-Cola participates in numerous community development programmes, with an emphasis on education, health, and communal infrastructure. They provide funding for initiatives that improve education, expand access to healthcare, and foster long-term community development.

Coca-Cola participates in numerous community development initiatives, with an emphasis on education, health, and community infrastructure. They provide funding for initiatives that improve education, expand access to healthcare, and foster long-term community development.

7.8: Environmental Stewardship

Corporate social responsibility activities that emphasise environmental sustainability help to make the world a better place. Companies that apply sustainable practises, save resources, reduce carbon emissions, and promote renewable energy can help to minimise climate change and protect natural ecosystems.

Sustainable Operations	CSR initiatives exhort businesses to implement sustainable business practises. This includes reducing energy consumption, eliminating waste, and instituting eco-friendly technologies and practises. By

	employing sustainable business practises, businesses can reduce their environmental impact and contribute to environmental stewardship.
Natural Resource Conservation	Conservation of natural resources, such as water, forests, and biodiversity, is a common focus of corporate social responsibility initiatives. Companies can reduce water consumption, promote responsible raw material procurement, and support conservation efforts by taking the appropriate measures. These initiatives contribute to the long-term preservation and management of natural resources for future generations.
Renewable Energy Adoption	Numerous CSR initiatives emphasise the transition to renewable energy sources. Investing in renewable energy infrastructure, installing solar panels or wind generators, or purchasing renewable energy credits are options for businesses. By switching to clean and renewable energy sources, businesses reduce their reliance on fossil fuels and contribute to climate change mitigation and air pollution reduction.
Carbon Footprint Reduction	Commonly included in CSR programmes are goals for reducing greenhouse gas emissions and carbon footprints. Carbon emissions can be offset by reforestation or renewable energy initiatives, as well as by

   The Comprehensive Guide On CSR For Consultant

	implementing energy-efficient measures and promoting sustainable transportation options. Companies demonstrate their commitment to environmental stewardship and contribute to global climate action by actively reducing their carbon footprint.
Environmental Education and Awareness	CSR projects that promote environmental education and develop environmental awareness can be included in CSR programmes. Companies can educate employees, customers, and local communities about environmental challenges, sustainable practises, and the value of environmental stewardship. Companies that promote environmental literacy allow individuals to make more sustainable choices and contribute to a greener future.
Supply Chain Sustainability	Supply chain management is commonly incorporated into CSR programmes. Businesses can collaborate with their suppliers to promote sustainable procurement practises, uphold ethical labour standards, and reduce environmental impacts throughout the supply chain. By promoting sustainability within their supply networks, companies can generate positive environmental change beyond their own operations.

 The Comprehensive Guide On CSR For Consultant

Example

Interface: Interface, a global carpet tile manufacturing company, is widely recognized for its commitment to sustainability and environmental stewardship. They have implemented several CSR programs that prioritize environmental protection:

Mission Zero: Interface has set a bold sustainability goal called "Mission Zero," which aims to eliminate any negative impact their operations have on the environment by 2020. They have made significant progress in reducing carbon emissions, waste, and water usage throughout their manufacturing process.

Sustainable Materials: Interface is committed to sourcing sustainable materials for their products. They pioneered the use of recycled materials in carpet tiles and have developed innovative techniques to recycle and reuse old carpets, reducing waste sent to landfills.

Climate Take Back: Interface's Climate Take Back initiative goes beyond reducing their carbon footprint. It focuses on regenerative practices that aim to reverse climate change. They invest in renewable energy, carbon sequestration projects, and ecosystem restoration efforts to achieve a positive environmental impact.

IKEA:

IKEA, the global furniture retailer, has implemented various CSR programs to prioritize environmental stewardship:

 The Comprehensive Guide On CSR For Consultant

Sustainable Sourcing: IKEA is committed to sourcing materials from sustainable and responsibly managed sources. They strive to ensure that wood, cotton, and other raw materials used in their products come from renewable sources and meet strict environmental and social criteria.

Renewable Energy Investment: IKEA has invested heavily in renewable energy. They have installed solar panels on many of their stores and warehouses, allowing them to generate clean energy on-site. Additionally, they have invested in wind farms and purchased renewable energy certificates to offset their energy consumption.

Circular Economy Initiatives: IKEA is actively working towards a circular economy model. They promote product longevity and encourage customers to repair, reuse, and recycle their products. IKEA also offers a buy-back program where customers can return used furniture for a store credit, enabling the company to refurbish and resell the items.

The above examples illustrate how IKEA and Interface have incorporated environmental stewardship into their CSR initiatives. Through sustainable sourcing, investments in renewable energy, initiatives for a circular economy, and responsible production practises, these companies are actively working to reduce their environmental impact and contribute to a more sustainable future.

7.9: Regulatory Compliance and Public Relations

Strong CSR programmes provide legal compliance. By exceeding regulatory requirements, businesses might avoid penalties and bad publicity. Effective CSR programmes

 The Comprehensive Guide On CSR For Consultant

also help organisations demonstrate their social and environmental benefits, building stakeholder trust.

Businesses can exceed regulations through CSR programmes. By adopting ethical and sustainable practises, companies demonstrate their ESG compliance. This can help organisations avoid noncompliance penalties, legal issues, and reputational damage. CSR helps companies identify and mitigate risks. Ethical and sustainable practises can reduce environmental pollution, workplace abuses, supply chain challenges, and other ethical concerns. Proactive risk management reduces legal problems, regulatory transgressions, and negative press.

CSR programmes boost PR and media coverage. Social, environmental, and community-focused companies generate stories and content that may be shared on multiple channels. Positive media coverage enhances a company's brand, reputation, and social and environmental responsibility. ESG factors are being considered by investors. CSR programmes show a company's commitment to sustainability. This may attract ethical investors and improve investor relations. ESG-focused investment funds and indices may support ESG-focused enterprises.

Example

Patanjali Ayurved:

Patanjali Ayurved, an Indian consumer goods company known for its natural and Ayurvedic products, has leveraged CSR programs to ensure regulatory compliance and strengthen public relations:

   The Comprehensive Guide On CSR For Consultant

Quality Assurance and Standardization: Patanjali Ayurved focuses on maintaining high product quality and adhering to regulatory standards. They follow Good Manufacturing Practices (GMP) and invest in quality control measures to ensure that their products meet safety and efficacy requirements.

Social Welfare Initiatives: Patanjali Ayurved is involved in various social welfare programs aimed at benefiting communities and addressing societal needs. They have initiatives focused on education, healthcare, rural development, and promoting Indian traditional knowledge and heritage. These programs help build positive public perception and enhance their reputation as a socially responsible company.

Consumer Education: Patanjali Ayurved emphasizes consumer education and awareness through its marketing and communication campaigns. They provide information about the benefits of Ayurvedic products, natural ingredients, and the importance of sustainable living. This helps in building consumer trust and loyalty while complying with regulations related to product claims and advertising.

Nestlé:

Nestlé, a multinational food and beverage company, has implemented CSR programs to ensure regulatory compliance and enhance public relations:

Responsible Sourcing: Nestlé is committed to responsible sourcing of raw materials, especially in areas like cocoa, coffee, and palm oil. They have established stringent guidelines and partnerships with suppliers to ensure

 The Comprehensive Guide On CSR For Consultant

sustainable and ethical sourcing practices, addressing concerns related to deforestation, child labor, and human rights violations.

Product Labeling and Transparency: Nestlé is proactive in providing accurate and transparent information to consumers. They adhere to regulatory requirements for product labeling, nutritional information, and allergen disclosures. Nestlé also engages in educational campaigns to promote healthy eating habits and combat issues like obesity and malnutrition.

Stakeholder Engagement: Nestlé actively engages with various stakeholders, including NGOs, governments, and local communities, to address concerns and collaborate on sustainability initiatives. They conduct regular stakeholder consultations, participate in industry dialogues, and transparently report on their CSR efforts to foster trust and build positive relationships.

These examples show how corporations such as Nestlé and Patanjali Ayurved have used CSR programmes to secure regulatory compliance and improve public relations. These companies have strengthened their compliance efforts and built positive relationships with consumers, communities, and regulatory bodies by implementing responsible sourcing practises, ensuring product quality and transparency, engaging with stakeholders, and undertaking social welfare initiatives.

7.10: Improved Supply Chain Management

Strong CSR programmes ensure ethical sourcing, labour standards, and environmental compliance across supply chains.

 The Comprehensive Guide On CSR For Consultant

Corporate social responsibility projects support ethical sourcing in supply chains. Suppliers must be checked for labour, human rights, and environmental compliance. Ethical sourcing protects brands from unethical vendors.

CSR programmes include suppliers in sustainability and social responsibility. Suppliers can help companies promote ethical sourcing, environmental stewardship, and fair employment. Collaboration improves supply chain management by fostering teamwork and accountability.

CSR programmes value supply chain openness and traceability. Companies may trace and monitor raw commodities to ensure responsible and sustainable sourcing. Transparent supply chains help companies solve social and environmental challenges, building trust.

CSR programmes often analyse suppliers' performance. Companies may evaluate suppliers' responsible practises using benchmarks and indicators. This assessment process encourages suppliers to meet the company's sustainability goals, creating a more sustainable and responsible supply chain.

CSR programmes foster supplier-company collaboration for ongoing improvement. Sharing best practises, expertise, and resources with suppliers can promote supply chain sustainability and social responsibility. This partnership could lead to innovative solutions, improved efficiency, and less environmental and social impacts. CSR programmes can mitigate supply chain risks. Businesses can analyse and remedy human rights violations, child labour, unsafe working conditions, environmental impacts, and other ethical issues. Proactively addressing these concerns can strengthen supply chains.

Supply chain partners might be prioritised. Companies and major suppliers can enhance supply chain management by communicating, creating goals, and addressing sustainability issues.

Example

Walmart:

Walmart, a multinational retailer, has instituted CSR initiatives to improve supply chain management:

Walmart's Supplier Sustainability Programme focuses on promoting responsible procurement, environmental sustainability, and social compliance throughout its supply chain. They collaborate closely with suppliers to establish expectations, provide training and resources, and conduct audits to ensure adherence to ethical and sustainable practises.

Transparency and Traceability: In its supply chain, Walmart prioritises transparency and traceability. They require suppliers to disclose the origins of their products, including the basic materials and manufacturing processes. Walmart is piloting projects to enhance product traceability using technologies such as blockchain, enabling consumers to access information about the journey of products from farm to shelf.

Walmart encourages supplier collaboration in order to advance its sustainability initiatives. They engage in dialogue and partnerships with suppliers to identify enhancement opportunities, share best practises, and

 The Comprehensive Guide On CSR For Consultant

collaborate on sustainability initiatives. This collaborative strategy facilitates the development of stronger relationships and promotes positive change throughout the supply chain.

H&M:

Global fashion retailer H&M has implemented CSR initiatives to improve supply chain management:

Supplier Audits and Compliance: H&M has enacted stringent supplier compliance standards and conducts routine audits to ensure environmental, social, and labour requirements are met. They evaluate suppliers using standards such as fair wages, secure working conditions, and responsible chemical management. H&M identifies areas for improvement and collaborates with suppliers to implement corrective actions as a result of these audits.

Significant efforts have been made by H&M to implement sustainable materials into its supply chain and to recycle. They promote the use of organic cotton, recycled materials, and responsible raw material procurement. Through in-store recycling programmes, H&M also encourages customers to recycle their clothing in an effort to create a closed-loop system and reduce waste in the fashion industry.

Training and Capacity Building: H&M invests in training and capacity building programmes for its suppliers. They provide suppliers with support and resources to enhance their sustainability practises, such as water and energy management, waste reduction, and worker empowerment. These programmes strengthen supplier capabilities and promote continuous supply chain improvement.

 The Comprehensive Guide On CSR For Consultant

Intel:

Intel, a multinational technology firm, has incorporated CSR initiatives to improve supply chain management:

Intel has taken proactive measures to ensure responsible sourcing of minerals in its supply chain in compliance with the Conflict Minerals Act. They have instituted rigorous due diligence procedures to prevent the use of conflict minerals such as tin, tantalum, tungsten, and gold, which can be used to finance armed conflict and human rights violations. Intel collaborates with suppliers to establish mineral supply chain traceability and transparency.

Supplier Diversity and Inclusion: Intel is committed to diversity and inclusion among its suppliers. They actively endeavour to engage and support a diverse array of suppliers, including minority-, women-, and small-owned companies. Intel facilitates economic growth, supports local communities, and creates a more inclusive supply chain ecosystem by promoting supplier diversity.

Intel collaborates with its suppliers to enhance environmental performance. They establish environmental goals and engage in regular dialogues with their suppliers to share best practises, identify areas for development, and foster innovation in areas such as energy efficiency, waste reduction, and water conservation.

These examples illustrate how Walmart, H&M, and Intel have utilised CSR initiatives to enhance supply chain management. Through supplier compliance, sustainable materials, capacity development, conflict minerals compliance, supplier diversity, environmental performance, and collaborative activities, these businesses have

 The Comprehensive Guide On CSR For Consultant

improved the transparency, sustainability, and ethical practises of their supply chains.

7.11: Long-Term Business Viability

CSR initiatives consider social, environmental, and economic concerns to ensure long-term viability and success.

CSR activities help the company build a good reputation by showing its social and environmental responsibility. The company's reputation benefits customers, employees, investors, and other stakeholders who share its values. Good reputation and brand value increase consumer loyalty, top talent, and long-term investment, ensuring firm viability. CSR programmes engage customers, employees, communities, and regulators. Involving stakeholders in decision-making and addressing their concerns builds trust, loyalty, and long-term partnerships. Positive stakeholder relations help the firm thrive.

CSR programmes may distinguish businesses. Socially and environmentally responsible companies stand out from competition and attract clients that appreciate sustainable and ethical products and services. This market distinctiveness ensures long-term economic viability by attracting customers and market share.

Example

Interface Inc.:

Interface Inc., a global modular flooring manufacturer, has implemented CSR programs to enhance long-term business viability:

   The Comprehensive Guide On CSR For Consultant

Mission Zero: Interface is known for its ambitious sustainability program called "Mission Zero." The company aims to eliminate any negative impact on the environment by 2030. Through this program, Interface has implemented various initiatives such as reducing greenhouse gas emissions, increasing energy efficiency, promoting recycling, and sourcing sustainable materials. These efforts have not only reduced the company's environmental footprint but also positioned Interface as a leader in sustainable flooring solutions, attracting environmentally conscious customers and driving long-term business growth.

Circular Economy Approach: Interface has embraced a circular economy approach, focusing on designing products that can be recycled and repurposed at the end of their life cycle. The company actively engages with customers to take back used flooring materials and recycle them into new products. By adopting a closed-loop system, Interface reduces waste, conserves resources, and builds a more sustainable business model that aligns with changing customer preferences for sustainable products.

Danone:

Danone, a multinational food products company, has implemented CSR programs to enhance long-term business viability:

B Corp Certification: Danone has achieved B Corp certification, which verifies the company's commitment to meeting high standards of social and environmental performance, transparency, and accountability. This certification demonstrates Danone's long-term commitment

 The Comprehensive Guide On CSR For Consultant

to sustainable business practices and responsible corporate citizenship. It helps build trust with stakeholders, attracts socially conscious consumers, and differentiates Danone from competitors.

Sustainable Sourcing and Agriculture: Danone has made significant efforts to promote sustainable sourcing and agriculture practices. They work closely with farmers to implement sustainable farming methods, reduce carbon emissions, and protect biodiversity. By supporting regenerative agriculture and responsible sourcing, Danone ensures the long-term availability of quality ingredients, reduces supply chain risks, and contributes to the resilience of local farming communities.

Social Impact Initiatives: Danone has launched social impact initiatives to address pressing social and environmental challenges. For example, they have implemented programs to improve access to clean water, promote nutrition education, and support community development in underprivileged areas. These initiatives not only have a positive social impact but also contribute to long-term business viability by strengthening Danone's reputation, building brand loyalty, and creating shared value.

(**B. Corp Certification:** B Corp Certification is an award given to businesses that meet high social and environmental performance, transparency, and accountability requirements. B Corps, or Benefit Corporations, are for-profit businesses that have made the commitment to use their business as a force for good and to have a beneficial impact on society and the environment. A corporation must pass a thorough review undertaken by the nonprofit

   The Comprehensive Guide On CSR For Consultant

organisation B Lab to become a certified B Corp. This evaluation assesses the company's performance in a variety of categories, including governance, workers, community, environment, and customers. The evaluation considers aspects such as the company's mission, the impact on stakeholders, sustainability practises, and ethical business practises.

The B Corp Certification is a continuous commitment to maintaining and increasing social and environmental performance, rather than a one-time recognition. Certified B Corps must meet certain performance and transparency standards and must be recertified every three years to ensure they continue to meet the standards.)

These examples show how organisations such as Interface Inc. and Danone have used CSR programmes to build long-term financial viability. These companies have demonstrated their commitment to long-term sustainable growth, resilience, and positive social and environmental impacts through ambitious sustainability goals, circular economy approaches, B Corp certification, sustainable sourcing, agriculture initiatives, and social impact programmes.

7.12: Contributing to Sustainable Development Goals (SDGs)

Robust CSR programmes help achieve the UN Sustainable Development Goals.

 CSR and sustainability programmes let companies align their efforts with industry-specific SDGs. Selecting the SDGs most relevant to their business allows companies to focus their efforts and resources on improving them.

   The Comprehensive Guide On CSR For Consultant

Businesses can implement major SDG actions through CSR and sustainability programmes. Operations, supply chains, and community participation can help companies meet SDG targets like poverty reduction, good education, clean energy, sustainable consumption and production, and more. These actions advance SDGs.

CSR and sustainability programmes sometimes require partnership with governments, NGOs, local communities, and other businesses. Partnerships and shared sustainability goals can help companies impact the SDGs. Sharing information, resources, and best practises improves solutions and SDG progress.

CSR and sustainability projects drive innovation and sustainable technology. R&D investments support SDG goals like innovation, sustainable infrastructure, and climate action. Sustainable innovations can boost growth, quality of life, and progress. CSR and sustainability programmes require supply chain ethics. Suppliers can encourage decent labour, human rights, and responsible sourcing by following social and environmental standards. Sustainable supply chains support SDG goals for decent work, economic growth, responsible consumption and production, and reduced inequities.

Sustainability and CSR highlight environmental sustainability. Companies reduce carbon emissions, conserve natural resources, manage waste, and protect biodiversity to meet SDG targets for climate change, clean energy, responsible consumption and production, and life below and above water.

Case study:

   The Comprehensive Guide On CSR For Consultant

GSK (GlaxoSmithKline):

GSK, a global pharmaceutical company, has integrated the SDGs into its CSR initiatives:

Access to Healthcare: GSK is committed to improving access to healthcare, particularly in low-income countries. They have implemented various programs to provide affordable medicines, vaccines, and healthcare services to underserved populations. This effort aligns with SDG 3 (Good Health and Well-being).

Sustainable Operations: GSK focuses on reducing its environmental impact and promoting sustainable operations. They have set ambitious targets to reduce greenhouse gas emissions, water consumption, and waste generation across their value chain. This contributes to SDG 12 (Responsible Consumption and Production) and SDG 13 (Climate Action).

Example :

Siemens:

Siemens, a global technology company, has integrated the SDGs into its CSR initiatives:

Sustainable Cities: Siemens has implemented projects focused on creating smart and sustainable cities. They provide solutions for efficient energy management, intelligent transportation systems, and sustainable infrastructure development. This aligns with SDG 11 (Sustainable Cities and Communities).

Climate Action: Siemens has a strong commitment to combating climate change. They develop and implement

 The Comprehensive Guide On CSR For Consultant

technologies for renewable energy generation, energy efficiency, and emissions reduction. This contributes to SDG 7 (Affordable and Clean Energy) and SDG 13 (Climate Action).

Unilever:

Unilever, a multinational consumer goods company, has integrated the SDGs into its CSR initiatives:

Sustainable Sourcing: Unilever is committed to sustainable sourcing of raw materials. They work closely with farmers, smallholders, and suppliers to promote sustainable agricultural practices, protect biodiversity, and improve livelihoods. This effort aligns with SDG 2 (Zero Hunger) and SDG 15 (Life on Land).

Health and Hygiene: Unilever has launched various initiatives to improve health and hygiene practices globally. Their programs focus on providing access to clean water, promoting handwashing, and raising awareness about sanitation. This contributes to SDG 3 (Good Health and Well-being) and SDG 6 (Clean Water and Sanitation).

Microsoft:

Microsoft, a global technology company, has incorporated the SDGs into its CSR agenda:

Digital Inclusion: Microsoft is committed to bridging the digital divide and ensuring digital inclusion for all. They provide technology access, digital skills training, and resources to underserved communities, contributing to SDG 4 (Quality Education) and SDG 9 (Industry, Innovation, and Infrastructure).

    The Comprehensive Guide On CSR For Consultant

Environmental Sustainability: Microsoft has set ambitious environmental targets, including becoming carbon negative by 2030. They invest in renewable energy projects, promote energy efficiency, and work towards responsible waste management. This aligns with SDG 7 (Affordable and Clean Energy) and SDG 13 (Climate Action).

Coca-Cola:

Coca-Cola, a global beverage company, has aligned its CSR programs with the SDGs:

Water Stewardship: Coca-Cola has made significant efforts to conserve water resources and improve water access in water-stressed areas. They have implemented water replenishment projects, collaborated with local communities, and promoted water conservation practices. This effort aligns with SDG 6 (Clean Water and Sanitation).

Women's Empowerment: Coca-Cola has prioritized women's empowerment through various initiatives. They have focused on promoting women's economic empowerment, entrepreneurship, and leadership development. These efforts contribute to SDG 5 (Gender Equality) and SDG 8 (Decent Work and Economic Growth).

IKEA:

IKEA, a global furniture retailer, has embraced sustainability and social responsibility:

Renewable Energy: IKEA has invested heavily in renewable energy sources such as wind and solar. They aim to produce as much renewable energy as they consume in

 The Comprehensive Guide On CSR For Consultant

their operations and have installed solar panels on their stores and distribution centers. This effort contributes to SDG 7 (Affordable and Clean Energy).

Responsible Sourcing and Production: IKEA has implemented sustainable sourcing practices, including the use of responsibly harvested wood and the promotion of fair labor standards in their supply chain. They also focus on designing products with a circular economy approach, encouraging recycling and minimizing waste. This aligns with SDG 12 (Responsible Consumption and Production).

These examples show how businesses have integrated CSR programmes to contribute to the SDGs. These companies are creating a beneficial influence on society and the environment while promoting sustainable development by aligning their strategies, operations, and projects with specified aims.

7.13: Improved Governance and Transparency

CSR programmes encourage ethical behaviour, transparency, and accountability.

Accountability: CSR and sustainability programmes increase organisational accountability. Companies set social and environmental performance goals and targets to evaluate and report success. This promotes accountability and openness.

Stakeholder engagement: CSR and sustainability programmes encourage stakeholders like employees,

 The Comprehensive Guide On CSR For Consultant

consumers, communities, investors, and regulators to engage. By include stakeholders in decision-making and asking feedback, companies may ensure that multiple perspectives are acknowledged and integrated into governance. Participation improves transparency, trust, and decision-making.

Standards and Guidelines: CSR and sustainability programmes frequently follow international standards, guidelines, and reporting structures. Companies exhibit excellent governance, transparency, and accountability by voluntarily embracing such norms. These standards boost credibility and allow industry comparisons.

Transparency: CSR and sustainability programmes demand corporations to publish their social, environmental, and economic performance. Transparent reporting helps stakeholders understand the company's activities and impacts. Companies demonstrate transparency and empower stakeholders with accurate and complete reporting.

Board Oversight and Responsibility: CSR and Sustainability initiatives encourage board oversight of social and environmental challenges. The board sets the company's strategic direction, including its CSR and environmental goals. Sustainability in board conversations and decision-making improves governance and long-term wealth development.

Whistleblowing Mechanisms: CSR and sustainability plans usually require employees and stakeholders to report unethical behaviour, misconduct, and noncompliance. Whistleblower mechanisms enable anonymous complaints.

 The Comprehensive Guide On CSR For Consultant

This fosters organisational transparency, accountability, and good governance.

Example

Tata Group:

Tata Group, an Indian multinational conglomerate, has embraced governance and transparency as core values in its CSR efforts:

Ethical Conduct: Tata Group places strong emphasis on ethical conduct and responsible business practices. They have established comprehensive governance Structures, including codes of conduct, whistleblower mechanisms, and independent audits, to ensure transparency and integrity across their diverse business sectors. This aligns with SDG 16 (Peace, Justice, and Strong Institutions).

Corporate Philanthropy: Tata Group is actively involved in social development through its philanthropic initiatives. They support various social causes, such as education, healthcare, and environmental conservation, and maintain transparency in their philanthropic activities by providing regular reports on their impact and progress. This contributes to SDG 17 (Partnerships for the Goals).

Novartis:

Novartis, a global pharmaceutical company, has integrated governance and transparency into its CSR initiatives:

Ethics and Compliance: Novartis has implemented robust ethical standards and compliance measures across its operations. They have established strong governance Structures and policies to ensure transparency,

 The Comprehensive Guide On CSR For Consultant

accountability, and ethical conduct. This contributes to SDG 16 (Peace, Justice, and Strong Institutions).

Access to Medicines: Novartis has taken steps to improve access to essential medicines, particularly in low-income countries. They have implemented differential pricing strategies, partnered with local healthcare providers, and supported healthcare capacity-building initiatives. This effort aligns with SDG 3 (Good Health and Well-being).

Starbucks:

Starbucks, a multinational coffeehouse chain, has demonstrated a commitment to governance and transparency:

Ethical Sourcing: Starbucks focuses on responsible sourcing of coffee beans and other raw materials. They have established ethical sourcing standards, including fair trade practices, environmental sustainability, and supporting farmer welfare. This aligns with SDG 12 (Responsible Consumption and Production).

Transparency in Supply Chain: Starbucks promotes transparency in its supply chain, providing information about the origins of its products and the efforts taken to ensure sustainability. They engage in dialogue with stakeholders and regularly report on their sustainability performance. This contributes to SDG 17 (Partnerships for the Goals).

Nestlé:

Nestlé, a global food and beverage company, has prioritized governance and transparency in its CSR initiatives:

 The Comprehensive Guide On CSR For Consultant

Responsible Sourcing: Nestlé has implemented rigorous responsible sourcing practices for its agricultural commodities. They work closely with farmers and suppliers to ensure sustainable farming practices, protect human rights, and promote transparency in the supply chain. This effort aligns with SDG 12 (Responsible Consumption and Production).

Nutrition and Health Commitments: Nestlé has made commitments to improve the nutritional profile of its products and provide transparent and accurate information to consumers. They have implemented initiatives to reduce salt, sugar, and artificial ingredients in their products and have made efforts to promote healthier lifestyles. This contributes to SDG 3 (Good Health and Well-being).

These examples illustrate how corporations have integrated governance and openness into their CSR initiatives. By adopting high standards for responsible sourcing, ethical behaviour, and philanthropic transparency, these organisations contribute to enhanced governance, promote stakeholder trust, and drive positive social and environmental impact.

 The Comprehensive Guide On CSR For Consultant

Chapter 8: Indian Companies Act 2013 Clause 135

India pioneered CSR mandates for firms under Clause 135 of the Companies Act 2013. This rule mandated CSR spending by certain companies. The following factors have helped India dominate this field:

Government Initiative: Corporate responsibility and sustainability were recognised by the Indian government. CSR standards in the Companies Act were intended to encourage corporate responsibility and address social and environmental issues.

Corporate Accountability: India needed more corporate responsibility and openness. Campaigners, civil society organisations, and individuals pushed businesses to address poverty, education, healthcare, and environmental sustainability.

Public-Private Partnership: India's growth plan favours public-private partnerships. The government made CSR necessary to use the business sector's resources and talents to achieve societal goals.

Inequality and Economic Growth: Income inequality and societal inequities were highlighted by India's rapid economic expansion. CSR requirements closed the gap between corporate earnings and social development, supporting inclusive progress and shared prosperity.

The Comprehensive Guide On CSR For Consultant

Global Trends and Best Practises: India follows global corporate responsibility trends. India adopted the UN Global Compact and Sustainable Development Goals' views of CSR as a driver of sustainable development.

Stakeholder Engagement: Civil society, industry, and business executives actively participated in CSR discussions and dialogues. Their work shaped the Companies Act's CSR provisions.

Impact assessment and reporting: The Companies Act required enterprises to record their CSR activities, expenditures, and results. Due to greater openness and responsibility, companies created excellent CSR programmes.

Legal System: Clause 135 of the enterprises Act 2013, which superseded the Companies Act 1956, requires CSR for qualifying enterprises. This law established CSR as a legal mandate.

Rising Corporate Influence: Indian firms grew in size and influence, increasing their social and environmental impact. Increased corporate scandals and unethical practises prompted calls for increased corporate accountability and responsibility.

Focus on Inclusive Development: India seeks inclusive and sustainable growth. The government mandated CSR to use corporate resources to address social, economic, and environmental issues, particularly in marginalised and poor groups.

Indian mandated CSR has had significant social and environmental impacts. It has encouraged businesses to

 The Comprehensive Guide On CSR For Consultant

align their goals with sustainable development, encouraged CSR innovation, and enabled collaboration between businesses, governments, and civil society. India's required corporate social responsibility has inspired other nations to consider similar regulations.

Clause 135: Clause 135 of India's firms Act 2013 addresses Corporate Social Responsibility (CSR) rules for some firms. It requires corporations that meet specific conditions to spend a set percentage of their average net income on CSR efforts.

Here are some key points of Clause 135:

8.1: Applicability for Corporate Social Responsibility

Public Companies: All public companies, whether listed or unlisted, are subject to the CSR provisions of the Companies Act 2013 if they meet the financial criteria specified in the clause.

Private Companies: Private companies are also covered by Clause 135 if they meet the financial criteria mentioned in the provision.

Foreign Companies: Certain foreign companies that have a branch or project office in India and meet the specified financial criteria are also applicable under the CSR provisions.

Companies: Clause 135 applies to companies incorporated under the Companies Act, including public companies, private companies, and certain foreign companies.

- Financial Criteria: The provision applies to companies meeting any of the following criteria

 The Comprehensive Guide On CSR For Consultant

1. Companies with a net worth of **INR 500 crore** or more.
2. Companies with a turnover of **INR 1,000 crore** or more.
3. Companies with a net profit of **INR 5 crore** or more.

If a corporation hits any of these financial benchmarks in the previous fiscal year, it is obligated to comply with the CSR obligations and spend at least 2% of its average net profits produced in the three preceding fiscal years on CSR operations. CSR activities should be consistent with the list of qualifying activities outlined in the Companies Act of 2013. In addition, the company must establish a CSR Committee of the Board and report on its CSR activities and expenditures in its annual financial statements and Board report.

It's important to note that certain specific categories of companies are exempt from the CSR requirements, such as those formed as a charitable company under Section 8 of the Companies Act, companies with no or insufficient profits, and companies that are subsidiaries of other companies and meet certain conditions as per the Act.

8.2: Corporate Social Responsibility Expenditure

- Minimum Expenditure: Companies covered under Clause 135 are required to spend at least 2% of their average net profits made during the three immediately preceding financial years on CSR activities.
- Calculation of Net Profit: The net profit for the purpose of CSR expenditure calculation excludes certain specified items as defined in the Act.

Under Clause 135 of the Companies Act 2013 in India, there are no specific guidelines regarding the allocation or distribution of CSR expenditure across various activities. The Act mandates that companies with a specified net worth, turnover, or net profit must spend at least 2% of their average net profits made during the three immediately preceding financial years on CSR activities.

The Act does not prescribe a detailed breakdown of how the CSR expenditure should be allocated among different CSR activities. It gives companies the flexibility to choose CSR initiatives based on their business focus, expertise, and the needs of the communities they operate in.

While the Act does not provide specific guidelines for allocating CSR expenditure, it does require companies to formulate a CSR policy, approved by their Board of Directors, that outlines the nature of CSR activities they plan to undertake, the geographical areas of focus, and the implementation schedule. The CSR policy should guide the company in identifying and prioritizing CSR projects and initiatives.

The choice of CSR activities and their allocation of expenditure should be based on careful consideration of the social and environmental needs, potential impact, and long-term sustainability of the initiatives. Companies may conduct a needs assessment, engage with stakeholders, and prioritize projects that align with their core values and contribute to the well-being of society and the environment.

It's important for companies to be transparent and accountable in their CSR expenditure and report the details of their CSR activities in their annual financial statements

 The Comprehensive Guide On CSR For Consultant

and Board's report, as mandated by the Companies Act 2013.

8.3: Corporate Social Responsibility Thematic Area & Activities: The People, Planet, and Profit Structure, also known as the triple bottom line, is a concept that encourages companies to consider their social, environmental, and economic impacts when making business decisions.

Social

Here are some common corporate social responsibility (CSR) initiatives based on this Structure:

Education and Skill Development:

Education support: Providing scholarships, grants, or sponsorships for underprivileged students to access quality education.

Vocational training programs: Offering skill-building initiatives, apprenticeships, or vocational courses to enhance employability and empower individuals with relevant skills.

Literacy programs: Supporting initiatives that promote literacy and provide access to books, educational resources, or digital learning tools.

STEM education: Encouraging science, technology, engineering, and mathematics (STEM) education by funding programs or partnering with educational institutions.

 The Comprehensive Guide On CSR For Consultant

School infrastructure development: Contributing to the construction or renovation of schools and classrooms in underserved areas.

Health and Well-being:

Healthcare initiatives: Supporting healthcare infrastructure development, funding medical facilities, or sponsoring mobile clinics to provide healthcare services in underserved communities.

Disease prevention and awareness campaigns: Promoting awareness about diseases, supporting vaccination programs, and funding initiatives that focus on preventive healthcare.

Access to clean water and sanitation: Supporting projects that provide access to clean drinking water and improved sanitation facilities in areas with limited resources.

Nutrition programs: Addressing malnutrition and supporting initiatives that provide nutritious food, including school feeding programs or community gardens.

Maternal and child health: Supporting initiatives that focus on maternal and child healthcare, including prenatal care, immunizations, and access to proper nutrition.

Empowerment and Social Inclusion:

Women empowerment: Supporting programs that promote gender equality, economic empowerment, and entrepreneurship opportunities for women.

Youth development: Providing resources, mentorship programs, and skills training for youth to enhance their personal and professional development.

 The Comprehensive Guide On CSR For Consultant

Disability inclusion: Promoting inclusive employment practices, accessibility in infrastructure, and awareness campaigns to support individuals with disabilities.

Social entrepreneurship: Supporting social enterprises or startup incubators that address social challenges and create opportunities for marginalized communities.

Financial inclusion: Promoting access to financial services and supporting financial literacy initiatives to empower individuals and communities.

Community Development and Livelihood Enhancement:

Livelihood programs: Supporting income-generating activities, entrepreneurship training, or microfinance programs to help individuals and communities achieve sustainable livelihoods.

Infrastructure development: Contributing to the construction or improvement of infrastructure projects such as roads, bridges, community centers, or sanitation facilities.

Rural development: Supporting initiatives that address rural poverty, access to education, healthcare, and basic services in rural communities.

Disaster resilience and relief: Providing support during disasters, funding disaster preparedness programs, or contributing to emergency relief efforts in affected areas.

Cultural preservation: Supporting initiatives that preserve and promote local arts, crafts, traditions, and cultural heritage.

 The Comprehensive Guide On CSR For Consultant

Employee Volunteering and Skills Sharing:

Skills-based volunteering: Encouraging employees to use their professional skills and expertise to support nonprofits, social enterprises, or community organizations.

Mentoring programs: Establishing mentoring initiatives where employees provide guidance and support to individuals, such as students or aspiring professionals, to enhance their personal and professional development.

Pro bono services: Offering free services or expertise to nonprofit organizations or community projects in areas such as marketing, legal advice, technology, or business consulting.

Social Impact Partnerships:

Collaboration with NGOs: Establishing partnerships with nonprofit organizations to jointly address social issues and leverage resources, knowledge, and expertise.

Social innovation incubators: Supporting and funding programs or incubators that focus on social entrepreneurship, innovation, and the development of solutions for pressing societal challenges.

Collaboration with academia: Partnering with educational institutions to conduct research, develop solutions, or promote social initiatives through joint projects or programs.

Employee Well-being and Workforce Development:

Mental health support: Implementing employee assistance programs, providing resources, and promoting mental health awareness and support in the workplace.

 The Comprehensive Guide On CSR For Consultant

Workforce diversity and inclusion: Developing initiatives to foster diversity, equal opportunities, and inclusion within the workforce, such as diversity training, affinity groups, or ERGs.

Workforce development in disadvantaged communities: Creating employment opportunities, vocational training, or job placement programs in marginalized or economically challenged areas.

Employee well-being surveys and feedback mechanisms: Conducting regular surveys and feedback mechanisms to understand and address employees' well-being, satisfaction, and work-life balance.

Humanitarian Aid and Crisis Response:

Refugee support programs: Partnering with organizations that provide aid, resettlement support, and integration programs for refugees and displaced populations.

Humanitarian relief efforts: Providing financial aid, supplies, or volunteering support during humanitarian crises, natural disasters, or emergency situations.

Capacity-building initiatives: Supporting programs that enhance the capabilities and resilience of local communities, including disaster preparedness, vocational training, or income generation projects.

Responsible Marketing and Consumer Education:

Ethical advertising and marketing practices: Ensuring transparency, honesty, and responsible messaging in marketing campaigns, avoiding false claims or misleading information.

 The Comprehensive Guide On CSR For Consultant

Consumer education initiatives: Educating consumers about sustainable consumption practices, ethical choices, and the social or environmental impact of their purchasing decisions.

Product safety and customer welfare: Ensuring product safety, providing clear instructions, and actively addressing customer concerns or complaints in a timely manner.

Youth Education and Empowerment:

Scholarships and educational grants: Providing financial assistance to talented students from disadvantaged backgrounds to pursue higher education.

Career development programs: Offering mentorship, internship opportunities, or career guidance programs to help young individuals make informed career choices and develop necessary skills.

Digital literacy initiatives: Promoting access to technology and digital skills training to bridge the digital divide and empower youth in the digital age.

School infrastructure improvement: Investing in the construction, renovation, or maintenance of school buildings, classrooms, libraries, and educational facilities in underserved areas.

Education awareness campaigns: Conducting campaigns to raise awareness about the importance of education and increase enrollment rates in marginalized communities.

Gender Equality and Women's Empowerment:

Women's leadership programs: Establishing initiatives that provide mentoring, leadership training, and networking

 The Comprehensive Guide On CSR For Consultant

opportunities for women to advance their careers and leadership positions.

Financial inclusion for women: Supporting initiatives that promote access to financial services, microfinance, or entrepreneurship training specifically targeting women.

Combating gender-based violence: Partnering with organizations or funding programs that raise awareness, provide support, and promote prevention of gender-based violence.

Women's health and well-being: Supporting healthcare programs focused on women's reproductive health, maternal health, and access to healthcare services.

Advocacy for equal rights: Promoting policies and initiatives that strive for gender equality and equal opportunities in the workplace, including pay equity and flexible working arrangements.

Aging Population Support:

Elderly care programs: Supporting initiatives that provide healthcare, companionship, and social activities for elderly individuals, particularly those living alone or in care facilities.

Intergenerational programs: Facilitating connections between different generations through mentoring, educational programs, or community projects that promote understanding and cooperation.

Age-friendly infrastructure: Collaborating with local communities to develop infrastructure and services that

 The Comprehensive Guide On CSR For Consultant

cater to the needs of older adults, such as accessible transportation and public spaces.

Elderly employment initiatives: Creating opportunities for older adults to stay active and engaged through part-time employment, skills training, or entrepreneurship support.

Mental health support for seniors: Investing in programs that raise awareness about mental health issues affecting the elderly and providing access to mental health resources and services.

Indigenous Communities Support:

Cultural preservation initiatives: Supporting programs that preserve indigenous languages, traditional arts, crafts, and cultural heritage.

Economic empowerment: Collaborating with indigenous communities to develop sustainable livelihood opportunities, promote local entrepreneurship, or support fair trade initiatives.

Education and healthcare access: Establishing partnerships to improve access to quality education, healthcare, and social services in indigenous communities.

Land and resource rights advocacy: Supporting initiatives that advocate for the recognition and protection of indigenous land rights, resource management, and sustainable development practices.

Cultural exchange and awareness: Organizing events, workshops, or cultural exchange programs to foster understanding, appreciation, and respect for indigenous cultures.

 The Comprehensive Guide On CSR For Consultant

Refugee and Migrant Support:

Refugee integration programs: Partnering with organizations to provide support and resources for refugee integration, including language training, employment assistance, and social integration initiatives.

Legal aid and advocacy: Supporting legal aid programs that provide assistance to refugees and migrants, advocating for their rights, and promoting access to justice.

Access to basic needs: Providing food, shelter, healthcare, and other essential services to refugees and migrants in partnership with humanitarian organizations.

Education and skills training: Offering education and vocational training programs to enhance the skills and employability of refugees and migrants, facilitating their integration into local communities.

Community dialogue and cultural exchange: Promoting dialogue, understandingand cultural exchange between refugees, migrants, and local communities through events, workshops, and initiatives that foster mutual respect and collaboration.

Access to Basic Needs and Poverty Alleviation:

Food security programs: Supporting initiatives that address hunger and malnutrition by providing access to nutritious food, supporting food banks, or partnering with organizations focused on food security.

Affordable housing initiatives: Investing in affordable housing projects or partnering with organizations that

 The Comprehensive Guide On CSR For Consultant

provide safe and affordable housing options for low-income individuals and families.

Poverty alleviation programs: Supporting initiatives that focus on income generation, microfinance, and entrepreneurship training to empower individuals and communities to lift themselves out of poverty.

Social safety net support: Collaborating with organizations or government agencies to provide social assistance programs, such as cash transfer programs or healthcare subsidies, to vulnerable populations.

Access to clean energy: Supporting initiatives that promote affordable and clean energy solutions for underserved communities, such as renewable energy installations or clean cooking technologies.

Healthcare and Disease Prevention:

Health awareness campaigns: Conducting campaigns to raise awareness about prevalent health issues, preventive measures, and healthy lifestyle choices in collaboration with healthcare organizations or NGOs.

Healthcare infrastructure development: Investing in the construction or improvement of healthcare facilities, clinics, hospitals, or mobile medical units in underserved areas.

Access to healthcare services: Partnering with healthcare providers or organizations to offer free or subsidized healthcare services, medical check-ups, or vaccinations to disadvantaged communities.

 The Comprehensive Guide On CSR For Consultant

Disease prevention programs: Supporting initiatives focused on disease prevention, such as immunization campaigns, awareness programs, or initiatives targeting specific health issues like HIV/AIDS, malaria, or waterborne diseases.

Maternal and child healthcare: Promoting access to quality prenatal care, safe deliveries, postnatal support, and pediatric healthcare services for mothers and children.

Mental Health and Well-being:

Mental health awareness campaigns: Raising awareness about mental health issues, reducing stigma, and promoting access to mental health resources and support services.

Mental health training and education: Providing training programs or workshops to equip individuals, educators, and community leaders with the knowledge and skills to identify and address mental health challenges.

Support helplines and counseling services: Establishing helplines, online support platforms, or counseling services to offer immediate support and guidance to individuals experiencing mental health difficulties.

Workplace mental health programs: Implementing workplace initiatives that prioritize employee mental health, such as stress management programs, counseling services, or promoting work-life balance.

Collaborations with mental health organizations: Partnering with mental health organizations or nonprofits to support their initiatives, research, or community programs.

Humanitarian Aid and Disaster Response:

 The Comprehensive Guide On CSR For Consultant

Emergency relief efforts: Providing immediate assistance, supplies, and resources during natural disasters, conflicts, or humanitarian crises through partnerships with humanitarian organizations or through corporate philanthropic support.

Disaster preparedness and resilience programs: Supporting initiatives that focus on disaster preparedness, training, community resilience-building, and infrastructure development to mitigate the impact of disasters.

Refugee support and resettlement: Collaborating with organizations to provide support, resources, and integration assistance for refugees, including education, healthcare, livelihood opportunities, and psychosocial support.

Conflict resolution and peace-building: Investing in programs that promote peace, reconciliation, and conflict resolution in areas affected by conflict, including support for peace education, dialogue, and community-building initiatives.

Clean water and sanitation initiatives: Partnering with organizations to provide access to clean drinking water, sanitation facilities, and hygiene education to communities lacking basic water and sanitation infrastructure.

These CSR initiatives demonstrate a commitment to social development and addressing critical needs in society. By engaging in these activities, companies can contribute to improving the lives and well-being of individuals and communities, promoting social equity and sustainable development. I hope this list of corporate social responsibility initiatives based on the People Structure with

   The Comprehensive Guide On CSR For Consultant

social development activities provides you with further insight and ideas.

Environmental

Here are additional corporate social responsibility (CSR) initiatives based on the Planet Structure, with social and environmental development activities:

Renewable Energy Adoption and Climate Action:

Transition to renewable energy: Investing in renewable energy sources like solar or wind power to reduce greenhouse gas emissions and combat climate change.

Carbon footprint reduction: Implementing energy efficiency measures, adopting sustainable transportation solutions, and setting targets to reduce carbon emissions.

Climate advocacy and policy engagement: Supporting initiatives that advocate for strong climate policies, participating in global climate conferences, and collaborating with stakeholders to address climate-related challenges.

Sustainable transport initiatives: Promoting sustainable transportation options such as electric vehicles, public transportation, or carpooling to reduce carbon emissions and air pollution.

Deforestation prevention: Supporting programs that protect forests, combat deforestation, and promote sustainable land use practices.

Waste Management and Circular Economy:

 The Comprehensive Guide On CSR For Consultant

Waste reduction and recycling programs: Implementing waste management initiatives, promoting recycling practices, and supporting community-based recycling programs.

Product lifecycle assessment: Conducting assessments to minimize the environmental impact of products throughout their lifecycle, including design, production, use, and disposal.

Sustainable packaging solutions: Investing in eco-friendly packaging materials, promoting reusable packaging options, and reducing single-use plastics.

Circular economy initiatives: Embracing the principles of the circular economy by promoting resource efficiency, product repair, recycling, and upcycling practices.

E-waste management: Supporting responsible disposal and recycling of electronic waste, promoting e-waste collection programs, and partnering with recycling facilities.

Water Conservation and Protection:

Water stewardship programs: Implementing measures to conserve water resources, minimize water usage, and protect water quality in operations and supply chains.

Access to clean water: Supporting initiatives that provide access to clean and safe drinking water in underserved communities, including through the construction of water infrastructure or water purification systems.

Watershed restoration: Engaging in projects that restore and protect watersheds, wetlands, and water ecosystems to preserve biodiversity and ensure a sustainable water supply.

 The Comprehensive Guide On CSR For Consultant

Conservation education: Promoting awareness about the importance of water conservation and sustainable water management practices through educational programs and community outreach.

Rainwater harvesting initiatives: Supporting projects that promote rainwater harvesting techniques, such as rooftop rainwater collection systems or community water storage facilities.

Biodiversity Conservation and Ecosystem Protection:

Habitat restoration and conservation: Supporting projects that restore and protect habitats, including reforestation programs, wildlife conservation initiatives, and the establishment of protected areas.

Sustainable agriculture and land management: Encouraging sustainable farming practices, promoting organic agriculture, and supporting initiatives that conserve soil health and prevent land degradation.

Sustainable fisheries and marine conservation: Partnering with organizations or initiatives that promote sustainable fishing practices, protect marine ecosystems, and support the livelihoods of fishing communities.

Environmental education and awareness: Conducting educational campaigns and programs to raise awareness about biodiversity conservation, ecosystems, and the importance of preserving natural resources.

Ecosystem-based approaches: Promoting initiatives that prioritize ecosystem services and biodiversity, such as watershed management, coral reef restoration, or wildlife corridor preservation.

 The Comprehensive Guide On CSR For Consultant

Responsible Supply Chain Management:

Supply chain transparency: Implementing measures to ensure transparency and traceability within the supply chain, including audits, certifications, and responsible sourcing initiatives.

Supplier capacity building: Collaborating with suppliers to enhance their sustainability practices, provide training on social and environmental standards, and support their adoption of responsible business practices.

Ethical sourcing initiatives: Supporting initiatives that address issues like child labor, forced labor, human rights violations, and environmental degradation in supply chains.

Supplier diversity and inclusion: Promoting supplier diversity and inclusion by sourcing from minority-owned or women

Supplier diversity and inclusion: Promoting supplier diversity and inclusion by sourcing from minority-owned or women-owned businesses, fostering economic opportunities for underrepresented groups.

Sustainable procurement: Prioritizing the procurement of sustainable and eco-friendly products and services, considering factors such as environmental impact, social responsibility, and lifecycle assessment.

Ecosystem Restoration and Conservation:

 The Comprehensive Guide On CSR For Consultant

Land restoration projects: Engaging in initiatives that restore degraded lands, promote reforestation, and protect ecosystems to enhance biodiversity and ecosystem services.

Wetlands and coastal conservation: Supporting projects that protect and restore wetlands and coastal areas, which play a crucial role in carbon sequestration, water filtration, and wildlife habitats.

Conservation finance: Investing in funds or initiatives that provide financial support for conservation efforts, such as protected area management, wildlife conservation, or sustainable land use practices.

Collaboration with conservation organizations: Partnering with environmental NGOs or conservation groups to support their initiatives, fund research, or engage in joint conservation projects.

Green Technology Adoption and Innovation:

Research and development for sustainability: Investing in research and development to develop environmentally friendly technologies, sustainable materials, and energy-efficient solutions.

Green product design: Incorporating sustainability considerations into product design, such as using recyclable materials, reducing energy consumption, or minimizing waste generation.

Technology sharing and open innovation: Collaborating with other companies, academia, and organizations to share knowledge, resources, and technologies for sustainable development.

 The Comprehensive Guide On CSR For Consultant

Investment in clean technologies: Supporting startups or companies that develop clean technologies, renewable energy solutions, or innovations that contribute to environmental sustainability.

Technology for social impact: Promoting the use of technology to address social and environmental challenges, such as leveraging digital platforms for education, healthcare, or community development.

Water Stewardship and Conservation:

Watershed management: Participating in initiatives that protect and restore watersheds, promote water conservation, and ensure sustainable water use for communities and ecosystems.

Access to clean water: Supporting projects that provide access to clean drinking water and sanitation facilities in underserved areas, particularly in developing countries.

Water efficiency and conservation: Implementing water-saving measures in operations, promoting water conservation practices, and raising awareness about responsible water use among employees, customers, and communities.

Collaboration for water stewardship: Engaging in partnerships with local communities, governments, and NGOs to collectively manage water resources, address water-related challenges, and support water-dependent ecosystems.

Circular Economy and Waste Reduction:

 The Comprehensive Guide On CSR For Consultant

Product life cycle management: Designing products for durability, reparability, and recyclability, and promoting extended producer responsibility to minimize waste generation.

Circular business models: Exploring circular economy business models such as product leasing, repair services, or sharing platforms to maximize resource use and minimize waste.

Collaboration for waste management: Partnering with local communities, waste management organizations, or recycling initiatives to promote responsible waste management, recycling infrastructure development, and waste reduction campaigns.

Consumer education on waste reduction: Educating customers about responsible consumption, waste sorting, recycling practices, and the importance of reducing single-use products.

Upcycling and waste-to-energy initiatives: Supporting projects that convert waste into valuable resources, such as upcycling programs, waste-to-energy facilities, or composting initiatives.

Environmental Education and Awareness:

Environmental education programs: Supporting initiatives that provide environmental education and awareness in schools, communities, or through public campaigns to promote sustainable behaviors.

Conservation workshops and training: Organizing workshops and training sessions on environmental conservation, sustainable practices, and biodiversity protection for employees, communities, or youth.

Community engagement for environmental stewardship: Encouraging employees to participate in community clean-up activities, tree planting campaigns, or conservation projects to foster a sense of environmental responsibility.

Sustainable Agriculture and Food Systems:

Organic farming and regenerative agriculture: Promoting organic farming practices, regenerative agriculture methods, and agroecological approaches that enhance soil health, biodiversity, and reduce chemical inputs.

Support for smallholder farmers: Partnering with smallholder farmers, providing training, resources, and access to markets to enhance their productivity, income, and sustainability.

Food waste reduction: Implementing strategies to reduce food waste throughout the supply chain, such as food surplus redistribution, composting, or anaerobic digestion initiatives.

Sustainable sourcing of agricultural products: Ensuring the sourcing of agricultural products, such as cocoa, coffee, or palm oil, is done in a sustainable and responsible manner, avoiding deforestation and promoting fair trade practices.

Pollution Control and Remediation

Air pollution reduction: Implementing measures to reduce air emissions, such as upgrading equipment, adopting cleaner technologies, or promoting employee commuting alternatives.

Water pollution prevention: Implementing wastewater treatment systems, promoting water reuse and conservation, and adhering to strict effluent discharge standards to protect water quality.

Soil contamination remediation: Engaging in soil remediation initiatives, such as cleaning up contaminated sites, promoting sustainable land use practices, or supporting soil restoration projects.

Plastic pollution reduction: Implementing strategies to reduce single-use plastics, promoting plastic recycling and circular economy initiatives, and supporting plastic waste cleanup efforts.

Green Infrastructure and Urban Development

Sustainable building practices: Incorporating green building design principles, such as energy efficiency, water conservation, and use of eco-friendly materials, in construction projects.

Urban greening initiatives: Supporting urban tree planting programs, rooftop gardens, or green spaces that enhance biodiversity, mitigate heat island effects, and improve air quality in cities.

Sustainable transport infrastructure: Promoting the development of sustainable transport systems, such as bike lanes, electric vehicle charging stations, or public transportation networks to reduce carbon emissions and congestion.

   The Comprehensive Guide On CSR For Consultant

Smart cities and energy-efficient technologies: Investing in smart city solutions, energy-efficient lighting, or energy management systems to optimize resource use, reduce energy consumption, and enhance urban sustainability.

Stakeholder Engagement and Collaboration

Multi-stakeholder partnerships: Collaborating with governments, NGOs, communities, and industry peers to address shared environmental challenges, foster knowledge sharing, and implement joint initiatives.

Environmental impact assessments: Conducting thorough assessments of potential environmental impacts before initiating new projects, and implementing mitigation measures to minimize negative effects.

Transparency and reporting: Ensuring transparent reporting of environmental performance, including greenhouse gas emissions, water usage, waste generation, and progress towards sustainability goals.

Supplier sustainability engagement: Engaging suppliers in sustainability dialogues, providing guidelines, and encouraging them to adopt sustainable practices, reduce environmental impact, and promote social responsibility within their operations.

These CSR initiatives focus on the Planet Structure with social and environmental development activities, covering areas such as sustainable supply chain collaboration, ecosystem restoration, green technology adoption, water stewardship, and circular economy practices. By engaging in these initiatives, companies can contribute to the

    The Comprehensive Guide On CSR For Consultant

protection of the environment, promote sustainable practices, and drive positive social impact.

Financial

Here are corporate social responsibility (CSR) initiatives based on the Profit Structure, with activities that focus on social, environmental, and financial development:

Social Impact Investing:

Impact investing funds: Allocating funds toward investments that generate both social and financial returns, supporting social enterprises, and promoting sustainable development projects.

Venture philanthropy: Providing financial and non-financial support to social enterprises or nonprofit organizations with the goal of achieving social and financial sustainability.

Social impact bonds: Investing in social impact bond programs that link financial returns to the achievement of specific social outcomes, such as reducing recidivism rates or improving educational outcomes.

Employee Profit-Sharing and Ownership:

Employee stock ownership plans (ESOPs): Establishing programs that allow employees to own shares of the company, aligning their interests with the company's financial success.

Profit-sharing programs: Sharing a portion of company profits with employees through performance-based bonuses or profit distributions, fostering a sense of ownership and incentivizing productivity.

   The Comprehensive Guide On CSR For Consultant

Ethical Financial Practices:

Responsible lending and investing: Ensuring that financial products and services are offered in a responsible and ethical manner, avoiding predatory practices and investing in sustainable and socially responsible projects.

Financial education programs: Providing financial literacy programs and resources to empower individuals and communities to make informed financial decisions and build financial resilience.

Access to financial services: Expanding access to financial services in underserved communities, particularly low-income individuals and small businesses, through initiatives such as microfinance, mobile banking, or community banking.

Green Financing and Sustainable Development:

Green bonds: Issuing green bonds to finance environmentally sustainable projects, such as renewable energy installations, energy-efficient building projects, or sustainable infrastructure development.

Sustainable supply chain financing: Collaborating with suppliers to offer financing options that support their sustainability initiatives, such as eco-friendly production methods or responsible sourcing practices.

Environmental risk assessment: Incorporating environmental risk assessment in financial decision-making processes to identify and mitigate risks associated with climate change, natural resource depletion, or pollution.

Economic Empowerment and Job Creation:

 The Comprehensive Guide On CSR For Consultant

Entrepreneurship and small business support: Providing support to entrepreneurs and small businesses through mentorship programs, access to capital, and business development resources to foster economic growth and job creation.

Local sourcing and supplier development: Prioritizing local sourcing and partnering with local suppliers to support local economies, create employment opportunities, and build sustainable supply chains.

Skills development and vocational training: Investing in skills development programs that equip individuals with relevant skills for employment, entrepreneurship, and economic self-sufficiency.

Financial Transparency and Accountability:

Responsible reporting and governance: Ensuring transparency in financial reporting, adhering to accounting standards, and maintaining strong governance practices to build trust and credibility with stakeholders.

Stakeholder engagement on financial matters: Engaging with stakeholders, including shareholders, investors, and employees, to foster dialogue and provide information on financial performance, risks, and opportunities.

Impactful Product and Service Innovation:

Sustainable product development: Investing in research and development to create environmentally friendly products or services that reduce resource consumption, promote recycling, or have a positive social impact.

Accessible and inclusive product design: Ensuring that products and services are designed to be accessible to individuals with disabilities or special needs, promoting inclusivity and equal opportunities.

Affordable solutions for underserved markets: Developing products or services that are affordable and accessible to low-income or underserved populations, addressing social and financial inclusion.

Supplier Development and Responsible Sourcing:

Supplier capacity building: Collaborating with suppliers to enhance their sustainability practices, providing training, resources, and support to improve social and environmental performance.

Responsible supply chain management: Ensuring suppliers adhere to social and environmental standards, conducting audits, and supporting their efforts to improve responsible sourcing practices.

Supplier diversity and inclusion: Promoting diversity and inclusion within the supply chain by sourcing from minority-owned or women-owned businesses, fostering economic empowerment and resilience.

Stakeholder Engagement and Collaboration:

Multi-stakeholder partnerships: Collaborating with NGOs, local communities, governments, and industry peers to address shared challenges, leverage resources, and drive collective action on social and environmental issues.

Stakeholder consultation and feedback mechanisms: Establishing channels for stakeholders to provide input and

 The Comprehensive Guide On CSR For Consultant

feedback on company operations, products, and sustainability initiatives, fostering transparency and accountability.

Stakeholder-focused impact assessments: Conducting assessments to understand the social, environmental, and financial impacts of business activities on stakeholders, and developing strategies to mitigate negative impacts and maximize positive outcomes.

Ethical Business Practices and Governance:

Business ethics and anti-corruption measures: Implementing robust ethics policies, codes of conduct, and anti-corruption programs to promote integrity, transparency, and fair business practices.

Board diversity and governance practices: Fostering diverse and independent boards of directors, ensuring effective oversight of sustainability and long-term value creation.

Whistleblower protection: Establishing mechanisms to encourage employees and stakeholders to report unethical behavior, protecting whistleblowers from retaliation, and addressing reported concerns.

Impact Measurement and Reporting:

Integrated reporting: Adopting integrated reporting Structures that provide comprehensive disclosure of financial, social, and environmental performance, allowing stakeholders to assess the company's overall impact.

Key performance indicators (KPIs) for sustainability: Establishing measurable KPIs that track progress in social, environmental, and financial areas, enabling effective monitoring and continuous improvement.

Social return on investment (SROI) analysis: Assessing the social and environmental value created by the company's activities, projects, or initiatives, and reporting on the quantifiable social benefits delivered.

These CSR initiatives based on the Profit Structure demonstrate how companies can integrate social, environmental, and financial development activities to create shared value, drive sustainable business practices, and enhance long-term profitability.

However, the specific CSR activities a company chooses to undertake depend on its unique business focus, expertise, and the needs of the communities it serves. Each company can create its own CSR strategy and choose from a wide selection of socially beneficial projects that match with its core values and contribute to society's and the environment's well-being.

Companies must choose the most relevant and impactful CSR projects based on rigorous needs assessments, stakeholder discussions, and a clear grasp of the social and environmental concerns in their operating locations. Companies may maximise their positive impact and contribute to sustainable development by matching their CSR efforts with their business strengths and community needs.

8.4: Corporate Social Responsibility Committee

Under Clause 135 of the Companies Act 2013 in India, every company meeting certain financial criteria (e.g., net worth of INR 500 crore or more, turnover of INR 1,000 crore or more, or a net profit of INR 5 crore or more during any financial year) is required to constitute a CSR Committee of the Board. The CSR Committee is responsible for overseeing and approving the company's CSR policy and activities.

The guidelines for the CSR Committee under Clause 135 of the Companies Act 2013 are as follows:

Composition: The CSR Committee must consist of at least three directors, out of which at least one director should be an independent director. The Act requires the Board of Directors to appoint the members of the CSR Committee.

Responsibility of the CSR Committee:

In India, the CSR Committee of a company has various essential responsibilities related to the development, execution, and monitoring of the company's CSR activities under Clause 135 of the Companies Act 2013. The CSR Committee's primary responsibilities are as follows:

- **Formulate CSR Policy**: The CSR Committee is responsible for formulating and recommending the company's CSR policy. The policy should outline the company's approach to CSR, the areas of focus, and the activities to be undertaken. It should also specify the manner of executing these activities and the budget allocation for CSR initiatives.
- **Approve CSR Activities**: The CSR Committee is responsible for approving specific CSR projects and activities to be undertaken by the company. These

 The Comprehensive Guide On CSR For Consultant

activities should align with the CSR policy and contribute to the fulfillment of the company's CSR objectives.

- **Determine CSR Expenditure:** The Committee must ensure that the company spends at least 2% of its average net profits made during the three immediately preceding financial years on CSR activities. The committee should decide on the budget allocation for various CSR projects based on their potential impact and alignment with the CSR policy.

- **Monitor CSR Implementation:** The CSR Committee is responsible for monitoring the implementation of CSR activities to ensure that they are being carried out effectively and in line with the approved policy. Regular monitoring helps ensure that the company is making progress towards its CSR goals.

- **Assess Impact:** The Committee should assess the impact of the company's CSR initiatives on society and the environment. This involves evaluating the social, economic, and environmental outcomes of the CSR activities and measuring the positive changes they bring about.

- **Reporting:** The CSR Committee is required to report on its CSR activities in the Board's report, which forms part of the company's annual report. The report should provide details of the CSR projects undertaken, the amount spent on each project, and the outcomes achieved.

- **Align with Schedule VII:** The Committee should ensure that the company's CSR activities fall within

   The Comprehensive Guide On CSR For Consultant

the purview of Schedule VII of the Companies Act 2013, which specifies the broad categories of CSR activities that companies can undertake.

- **Periodic Review**: The CSR Committee may periodically review and update the CSR policy and activities based on changing circumstances, emerging needs, and feedback from stakeholders.
- **Engage with Stakeholders:** The Committee should engage with relevant stakeholders, including employees, communities, NGOs, and experts, to understand their needs and perspectives and to ensure that the company's CSR efforts address real societal and environmental challenges.
- **Ensure Compliance**: The CSR Committee must ensure that the company complies with all legal requirements related to CSR as per the Companies Act 2013 and the CSR Rules.

Overall, the CSR Committee plays a critical role in guiding and overseeing the company's CSR initiatives, ensuring that they are aligned with the company's values, contribute to sustainable development, and create a positive impact on society and the environment.

8.5: Corporate Social Responsibility Reporting

Companies in India are expected to follow specified requirements for reporting on their Corporate Social Responsibility (CSR) activities under Clause 135 of the Companies Act 2013. The following are the reporting guidelines for Clause 135:

CSR Report in Board's Report: The company must include a separate section on CSR in its Board's report,

 The Comprehensive Guide On CSR For Consultant

which forms part of the company's annual financial statements.

- **Disclosure of CSR Policy**: The report should include a brief outline of the company's CSR policy, which should include the activities undertaken, the areas of focus, and the rationale behind the chosen CSR initiatives.
- **CSR Expenditure**: The report should disclose the amount of money spent on CSR activities during the financial year. It should also specify the amount that was required to be spent as per the statutory requirement (2% of average net profits of the preceding three financial years) and the amount actually spent.
- **CSR Projects and Activities**: The company should provide a detailed description of the CSR projects and activities undertaken during the financial year. This includes information on the nature of the projects, the targeted beneficiaries, and the geographical locations where the projects were implemented.
- **Implementation Approach**: The report should describe the approach adopted by the company for implementing CSR initiatives and achieving the desired social and environmental outcomes.
- **Monitoring and Impact Assessment**: The company should disclose its monitoring and evaluation mechanisms for CSR projects. This includes the methods used to assess the impact of CSR initiatives on the intended beneficiaries and the community at large.

The Comprehensive Guide On CSR For Consultant

- **Challenges and Mitigation Strategies:** The report should highlight any challenges faced during the implementation of CSR projects and the strategies employed to address them.
- **Collaborations and Partnerships**: The company should disclose any collaborations, partnerships, or joint initiatives undertaken with other entities or organizations to implement CSR projects.
- **CSR Policy Review**: If the company has made any changes to its CSR policy during the financial year, the report should provide details of the amendments and the reasons behind them.
- **Non-compliance Explanation**: In case the company fails to spend the required amount on CSR or is unable to implement its CSR policy, the report should provide a detailed explanation for the non-compliance.

These rules are intended to provide openness and accountability in CSR reporting, as well as to offer stakeholders with a thorough understanding of the company's CSR programmes, their impact, and compliance with the regulatory standards set forth in Clause 135 of the Companies Act 2013.

Examples of India's few best CSR practitioners' companies:

1. **Tata Group**: Tata companies have a long history of pioneering CSR efforts in various areas, including education, healthcare, rural development, and environmental sustainability.
2. **Infosys:** Infosys is renowned for its CSR activities focused on education, healthcare, and community

development, as well as initiatives to promote sustainability and environmental conservation.

3. **Reliance Industries Limited**: Reliance is actively engaged in CSR projects related to healthcare, education, rural development, and disaster response, making a significant impact on communities.

4. **Mahindra Group**: Mahindra Group's CSR initiatives include sustainable livelihoods, education, and community welfare, reflecting its commitment to social and environmental responsibility.

5. **Wipro:** Wipro is dedicated to CSR efforts in areas such as education, healthcare, and community development, fostering positive change in society.

6. **Hindustan Zinc Limited**: Hindustan Zinc is actively involved in CSR projects focused on education, healthcare, women empowerment, and environmental conservation, particularly in the regions where it operates.

7. **Adani Group**: The Adani Group is committed to CSR efforts in areas such as education, healthcare, sustainable livelihoods, and rural development, contributing significantly to the communities it serves.

8. **Dr. Reddy's Laboratories**: Dr. Reddy's engages in CSR activities related to healthcare access, education, and community development, with a strong focus on improving health outcomes.

9. **Larsen & Toubro Infotech (LTI)**: LTI's CSR initiatives include education, skill development, and environmental sustainability, reflecting its

 The Comprehensive Guide On CSR For Consultant

commitment to social and environmental well-being.

10. **HUL (Hindustan Unilever Foundation):** Apart from HUL, the Hindustan Unilever Foundation, the CSR arm of HUL, plays a vital role in water conservation, sanitation, and rural development initiatives.

11. **ITC Limited**: ITC is actively engaged in CSR activities related to sustainable agriculture, rural development, education, and healthcare, making significant contributions to the well-being of communities.

12. **HDFC Bank**: HDFC Bank's CSR initiatives focus on education, skill development, and rural livelihood enhancement, supporting social and economic development across the country.

13. **Marico Limited**: Marico's CSR efforts include projects in healthcare, education, and sustainable livelihoods, demonstrating a commitment to social impact and community welfare.

14. **Ambuja Cements**: Ambuja Cements is known for its CSR initiatives in areas like water management, health, education, and environmental sustainability, promoting sustainable development in communities around its plants.

15. **JSW Group**: JSW Group's CSR activities cover education, healthcare, rural development, and sports, aiming to uplift the lives of people in the areas where it operates.

16. **Titan Company Limited**: Titan is involved in CSR activities focused on education, skill development,

   The Comprehensive Guide On CSR For Consultant

healthcare, and community welfare, contributing to the betterment of society.

17. **Godrej Group**: The Godrej Group's CSR initiatives encompass education, healthcare, environment, and rural development, aiming to create a positive social and environmental impact.

18. **Tata Consultancy Services (TCS):** TCS engages in CSR projects related to education, skill development, and technology for social development, leveraging its expertise for societal welfare.

19. **Aditya Birla Group**: Aditya Birla Group's CSR efforts include education, healthcare, sustainable livelihoods, and community development, aligning with its commitment to inclusive growth.

20. **Wockhardt Limited:** Wockhardt is actively involved in CSR activities related to healthcare access, sanitation, and rural development, focusing on improving health and well-being in underserved areas.

These companies have demonstrated a strong commitment to CSR and sustainability, implementing impactful initiatives that address social, economic, and environmental challenges. It's important to note that the CSR landscape in India is dynamic, and there are many other companies actively contributing to CSR and sustainability efforts across the country. For the most up-to-date information, it is recommended to refer to recent reports and assessments on CSR practices in India.

   The Comprehensive Guide On CSR For Consultant

CHAPTER 9: CORPORATE SOCIAL RESPONSIBILITY: A FRAMEWORK FOR PROGRAMME DEVELOPMENT

Corporate social responsibility (CSR) programmes need to be carefully thought out and aligned with the company's ideals, business goals, and stakeholder expectations. Here are some things companies can do to make sure their CSR projects are well-designed and well-thought-out:

9.1: Conduct a Materiality Assessment

A materiality evaluation is needed to figure out which social, environmental, and economic issues are the most important to your business and its stakeholders and to put them in order of importance. This assessment helps companies create successful corporate social responsibility (CSR) projects that focus on the most important and important-to-the-world problems. Here's how to do a step-by-step review of what's important:

Step 1: Identify Stakeholders

 The Comprehensive Guide On CSR For Consultant

Identify and involve key stakeholders, including employees, consumers, suppliers, investors, local communities, non-governmental organisations (NGOs), and industry professionals. Their feedback will help us comprehend their opinions and expectations regarding CSR issues.

Step 2: Collect Information

Collect information and data on a variety of prospective CSR issues. This can be accomplished through the use of surveys, interviews, public consultations, industry studies, and sustainable benchmarks.

Step 3: Evaluate the Effects

Think about the good and bad effects each CSR problem could have on the business and its stakeholders. Think about how big, wide, and bad the effects will be.

Step 4: Identify the importance of CSR issues.

Prioritise the CSR issues according to their significance to both stakeholders and the organisation. Priority is given to issues with a significant impact and a high level of stakeholder concern.

Step 5: Validation and verification

Validate the findings with the relevant stakeholders and verify the veracity of the evaluation. Verify that the examination meets the stakeholders' expectations.

Step 6: Create a Materiality Matrix.

   The Comprehensive Guide On CSR For Consultant

Present the results in a Materiality Matrix that illustrates the most pressing CSR issues. According to the matrix, the issues are divided into four quadrants:

HIGH PRIORITY AND HIGH IMPACT Issue 1 Issue 2 Issue 3	HIGH PRIORITY AND LOW IMPACT Issue 1 Issue 2 Issue 3
LOW PRIORITY AND HIGH IMPACT Issue 1 Issue 2 Issue 3	LOW PRIORITY AND LOW IMPACT. Issue 1 Issue 2 Issue 3

Case study

Amazon:

Amazon, the largest e-commerce and technology company in the world, recognises the significance of social responsibility and sustainability. Materiality Assessments are conducted by the company to identify the most pertinent ESG issues for its business and stakeholders.

Amazon's Materiality Assessment requires engagement with a variety of stakeholders, such as customers, employees, suppliers, investors, policymakers, and communities. Amazon collects information on the social and environmental impacts of its operations and supply chain through questionnaires, interviews, and discussions.

Amazon has identified significant issues such as carbon emissions, waste reduction, labour practises, data privacy, and diversity and inclusion based on the assessment. The CSR programmes of the company are designed to resolve these issues. Amazon, for instance, has made substantial commitments to achieve net-zero carbon emissions by 2040 and invests in renewable energy initiatives to sustainably power its operations. Through various initiatives, the company is also committed to enhancing working conditions for employees and supporting diverse communities.

Apple:

Apple Inc., a leading technology corporation renowned for its products and services, prioritises environmental sustainability and social responsibility. Regular Materiality Assessments are conducted by the company in order to identify the main ESG issues that matter to its stakeholders and align with its business objectives.

Apple assesses the social and environmental impacts of its products and operations in collaboration with customers, employees, suppliers, investors, and environmental experts. This evaluation aids in identifying material concerns such as carbon footprint reduction, responsible material procurement, product recycling, labour practises, and data privacy.

According to the evaluation, Apple has established comprehensive CSR programmes. The company aspires to attain a net-zero carbon footprint for its supply chain and products, invests in renewable energy, and designs recyclable and environmentally sustainable products. Apple supports education initiatives and provides its supply chain employees with training and development opportunities.

Materiality Assessments provide Amazon and Apple with significant insights into the concerns and expectations of their stakeholders, enabling them to construct more effective CSR programmes. These analyses serve as the cornerstone of their commitment to sustainability, assisting businesses in creating positive social and environmental benefits while promoting their long-term viability.

9.2: Setting Specific Goals for CSR Initiatives

Developing effective corporate social responsibility (CSR) programmes requires the establishment of specific objectives. Here is a step-by-step approach to establishing specific goals for CSR initiatives:

Before defining objectives, conduct a materiality analysis to identify the most significant environmental, social, and governance (ESG) issues affecting your organisation and its stakeholders. This analysis will assist you in prioritising and concentrating on the most essential CSR activities.

Align with Company Values and Purpose	Ensure that the CSR objectives align with the fundamental values, mission, and long-term vision of your organisation. CSR initiatives should enhance the company's reputation as a responsible corporate citizen and complement its mission.
Use the SMART	Set specific objectives using the SMART (Specific, Measurable,

   The Comprehensive Guide On CSR For Consultant

Structure	Achievable, Relevant, and Time-Bound) framework. Each objective should be well-defined, measurable, attainable, and time-bound.
Engage Stakeholders	Include relevant stakeholders in the goal-setting process, including employees, customers, suppliers, local communities, and NGOs. Consider their perspectives, requirements, and expectations to ensure that the objectives address the most pressing concerns.
Prioritize Impact Areas	Determine the main areas in which your CSR efforts can have the greatest positive impact. Prioritise initiatives that align with your materiality assessment and can create meaningful change, whether it's reducing carbon emissions, promoting diversity and inclusion, or supporting local communities.
Establish Baseline Metrics	Establish baseline metrics to assess the present status of each CSR objective. This will help monitor progress over time and evaluate the initiatives' efficacy.
Quantify Targets	Set distinct and measurable objectives for each objective. For instance, if your objective is to reduce greenhouse gas emissions, you should specify the percentage reduction you hope to achieve by a certain date.
Integrate the Goals into Business	Integrate CSR objectives into your organization's overarching business strategy. Ensure that they are backed by

   The Comprehensive Guide On CSR For Consultant

Strategy	senior leadership and integrated into daily operations and decision-making processes.
Create Action Plans	Create action plans outlining the specific steps and activities necessary to attain each objective. Accordingly assign responsibilities and allocate resources.
Monitor and Report Progress	Regularly monitor the progress of each CSR objective and disclose the results to stakeholders in a transparent manner. Track the impact of the initiatives using key performance indicators (KPIs) and make adjustments as necessary.

By following these steps and involving all key stakeholders, your organisation can define clear, meaningful, and impactful goals for successful CSR programmes. Consider that excellent CSR activities not only benefit society and the environment, but also the long-term viability and reputation of the company.

Example

Exxon Mobil:

Exxon Mobil conducts a comprehensive evaluation of materiality to determine the most significant environmental and social issues, taking into consideration stakeholder concerns and global sustainability challenges. The company's CSR objectives are consistent with its

 The Comprehensive Guide On CSR For Consultant

fundamental corporate values and long-term strategy, with a focus on addressing climate change, energy efficiency, and community development issues. CSR programmes of ExxonMobil:

Exxon Mobil is committed to addressing climate change through various initiatives. The company invests in research and technology to reduce greenhouse gas emissions and enhance energy efficiency.

Exxon Mobil implements programmes to protect biodiversity and natural resources as part of its environmental stewardship. It engages in conservation, protection of wildlife, and ecosystem restoration initiatives.

Community Development: The corporation engages in social and community development projects, particularly in regions in which it operates. Beneficial to local communities, it supports education, health, and livelihood development initiatives.

Safety and Wellbeing of Employees: Exxon Mobil places a premium on the safety and wellbeing of its employees. The organisation implements stringent safety protocols and offers health and wellness programmes for its employees.

Transparency and Reporting: The company maintains transparency by reporting its CSR activities and performance in sustainability reports on a regular basis. It discloses its environmental impacts, community investments, and progress towards achieving its sustainability objectives.

Ford Motor:

   The Comprehensive Guide On CSR For Consultant

Ford Motor Company conducts a materiality analysis to determine the most important sustainability issues, such as reducing carbon emissions, enhancing road safety, and developing electric mobility. Ford Motor Company establishes concrete and quantifiable objectives, such as reducing water usage in its manufacturing plants, increasing production of electric vehicles, and enhancing supply chain labour practises. Ford Motor's CSR initiatives.

Ford Motor Company is committed to the development of sustainable mobility solutions. The company, which has been a pioneer in the advancement of electric vehicle technology, intends to make substantial investments in electric and autonomous vehicles.

Ford promotes road safety through initiatives such as safety research, awareness campaigns, and the incorporation of safety features into their vehicles.

Diversity and Inclusion: The organisation places a premium on diversity and inclusion, nurturing an inclusive work environment and promoting supplier diversity. Ford values diverse perspectives and backgrounds among its personnel and supply chain.

Ford focuses on human rights, labour standards, and environmental sustainability throughout its entire supply chain to ensure responsible procurement and supply chain practises.

Ford participates in community service initiatives to address societal issues. In communities where it operates, the corporation supports educational programmes, environmental conservation efforts, and disaster relief initiatives.

 The Comprehensive Guide On CSR For Consultant

Ford intends to transition to a circular economy model by emphasising recycling and reuse of materials in vehicle production, reducing waste, and minimising environmental impacts.

Exxon Mobil and Ford Motor have both incorporated corporate social responsibility into their business strategies, thereby positively impacting society and the environment. Their commitment to sustainability, transparency, and responsible practises distinguishes them as corporate social responsibility leaders.

Exxon Mobil and Ford Motor Company both demonstrate how businesses can plan and define specific CSR objectives that align with their fundamental beliefs, resolve tangible challenges, and contribute to long-term growth. By embracing stakeholder interaction and transparent reporting, these companies demonstrate their dedication to responsible business practises and positive social and environmental impact.

9.3: Engaging the Key Stakeholders

Engagement of stakeholders is a crucial step in the development of successful corporate social responsibility (CSR) programmes. Consider how Microsoft engages stakeholders in its CSR efforts:

Example

Microsoft Corporation

 The Comprehensive Guide On CSR For Consultant

Multi-Stakeholder Consultations: Microsoft engages in extensive multi-stakeholder consultations to understand the diverse perspectives and needs of its stakeholders. It collaborates with customers, employees, NGOs, government representatives, and industry experts to identify key social and environmental issues that require attention.

Partnerships and Collaborations: Microsoft actively seeks partnerships and collaborations with NGOs, non-profits, and governmental organizations to co-create and implement CSR initiatives. One such partnership is the AI for Earth program, where Microsoft works with environmental organizations to leverage artificial intelligence for sustainable solutions.

Global Advocacy and Policy Engagement: Microsoft engages in global advocacy and policy discussions related to digital inclusion, privacy, cybersecurity, and environmental sustainability. By actively participating in these discussions, the company ensures that its CSR efforts align with broader societal goals.

Listening Tours and Town Halls: Microsoft conducts listening tours and town halls with its employees and communities to understand their concerns and expectations. This helps the company design CSR programs that resonate with its workforce and local communities.

Impact Measurement and Reporting: Microsoft places a strong emphasis on measuring the impact of its CSR initiatives and transparently reporting the results. This approach ensures accountability and demonstrates the company's commitment to stakeholders.

     The Comprehensive Guide On CSR For Consultant

Supplier Engagement: Microsoft extends its CSR efforts to its supply chain by engaging with suppliers to promote responsible business practices, ethical sourcing, and sustainability. The company sets expectations for its suppliers to align with its CSR goals.

Diversity and Inclusion: Microsoft focuses on promoting diversity and inclusion both within the company and in its CSR initiatives. The company actively engages with underrepresented communities to ensure that its prog

Microsoft ensures that its CSR programmes are aligned with the needs and expectations of its stakeholders and are meant to have a positive and meaningful influence on society and the environment through these stakeholder engagement practises. This approach not only improves the efficacy of Microsoft's CSR activities, but it also boosts the company's reputation as a socially responsible and purpose-driven organisation.

9.4: Integrate with Business Strategy

Integrating business strategy with corporate social responsibility (CSR) programmes is essential for ensuring that CSR efforts align with the organization's core values and long-term objectives. Consider how Unilever combines its commercial and CSR planning strategies:

Unilever:

Sustainable Living Plan: Unilever's Sustainable Living Plan (USLP) is at the core of the company's business strategy and CSR efforts. The USLP is an ambitious blueprint that aims to decouple the company's growth from its environmental footprint while increasing its positive social

   The Comprehensive Guide On CSR For Consultant

impact. It outlines specific targets and goals across various areas, such as reducing greenhouse gas emissions, improving water usage, enhancing livelihoods, and promoting sustainable sourcing of raw materials.

Business Model Transformation: Unilever has integrated sustainability into its business model by focusing on sustainable innovation and product development. The company has introduced numerous sustainable product lines, such as eco-friendly cleaning products, plant-based foods, and water-saving beauty products. This not only addresses consumer demands for sustainable products but also drives innovation and market growth.

Purpose-Driven Branding: Unilever's brands are aligned with the company's purpose of making sustainable living commonplace. For instance, the **Dove Self-Esteem** Project aims to promote body positivity and self-confidence among young people, while Lifebuoy's **"Help a Child Reach 5"** campaign focuses on promoting handwashing and reducing child mortality. These purpose-driven campaigns contribute to both brand reputation and positive social impact.

Stakeholder Engagement: Unilever actively engages with various stakeholders, including suppliers, consumers, NGOs, and governments. By involving stakeholders in its CSR planning and decision-making processes, the company ensures that its initiatives are well-aligned with societal needs and expectations.

Impact Measurement and Reporting: Unilever is committed to measuring the impact of its CSR initiatives and regularly publishes detailed sustainability reports. This transparent reporting demonstrates the company's progress toward its

 The Comprehensive Guide On CSR For Consultant

sustainability goals and fosters accountability to stakeholders.

Inclusive Business Models: Unilever embraces inclusive business models to address social challenges while creating business opportunities. For example, the company's Shakti program empowers women in rural India by training them as entrepreneurs to distribute Unilever products in their communities.

Supply Chain Sustainability: Unilever works closely with its suppliers to promote responsible sourcing, fair labor practices, and environmental stewardship. The company's Sustainable Agriculture Code sets standards for sustainable farming practices across its supply chain.

Unilever demonstrates that sustainability is a major component of how the company operates by incorporating CSR into its business strategy. This integration has not only boosted Unilever's reputation as a responsible corporate citizen, but it has also fueled innovation, strengthened brand loyalty, and contributed to long-term financial success.

9.5: Allocate Resources

Determine the financial, human, and technology resources required to effectively undertake CSR programmes. Adequate resource allocation is critical for CSR programme success.

Set Measurable Targets	Establish measurable and specific objectives for each CSR initiative. These objectives should be time-bound and consistent with the

	organization's overall CSR goals. Measurable objectives will aid in evaluating the impact of the initiatives and will direct resource allocation accordingly.
Assess Budget and Resources	Evaluate the company's CSR budget and the available resources. Consider allocating a specific portion of the budget to CSR initiatives. In addition, evaluate the availability of the human resources, skills, and expertise required for the implementation and management of CSR programmes.
Prioritize High-Impact Initiatives	Prioritise CSR initiatives that have the potential to have a significant positive impact, based on the materiality analysis and the defined objectives. Concentrate on projects that align with the company's primary competencies and where it can make a significant impact.
Consider Long-Term Sustainability	Ensure that the allocation of resources is long-term sustainable. Long-term planning enables the organisation to maintain continuity in its CSR efforts and establish enduring relationships with its stakeholders.

By referring to these steps, businesses can allocate resources strategically to their CSR programmes, resulting in more effective, sustainable, and fruitful initiatives.

Collaboration with others: Collaboration is essential for the design of effective corporate social responsibility

(CSR) programmes because it enables organisations to leverage diverse knowledge, resources, and perspectives for a greater positive impact. Here is an example of how businesses might collaborate with others to successfully plan CSR

Identify Relevant Stakeholders: Identify essential stakeholders with a vested interest in the CSR initiatives of the company. These parties may include local communities, non-governmental organisations, government agencies, suppliers, consumers, and industry associations.

Build Collaborative Partnerships: Develop collaborative relationships with pertinent stakeholders. Engage in transparent conversations and construct relationships based on mutual trust and esteem. Understand their CSR programme requirements and expectations.

Co-Create Solutions: Collaborate with stakeholders to co-develop innovative and sustainable solutions to the identified problems. To ensure inclusivity and ownership, include them in the planning, implementation, and evaluation of CSR programmes.

Pool Resources: Together with partners, pool financial, human, and technical resources to maximise the impact of CSR initiatives. Compared to isolated endeavours, sharing resources can result in more significant and transformative outcomes.

Engage in Collective Impact Initiatives: Participate in collective impact initiatives in which multiple stakeholders collaborate to achieve a common objective. These initiatives can bring about systemic change and more effectively address complex social issues.

 The Comprehensive Guide On CSR For Consultant

Cross-Sector Partnerships: Collaborate with organizations from different sectors, including non-profits, governments, and academic institutions. Each sector brings unique perspectives and expertise to address various aspects of CSR challenges.

Philanthropic Partnerships: Establish philanthropic partnerships with non-profit organizations to support community development and social projects. These partnerships can have a far-reaching impact on the lives of vulnerable populations.

Public-Private Partnerships: Engage in public-private partnerships (PPPs) with government entities to address social and environmental challenges. PPPs can combine the strengths of both sectors and mobilize resources for large-scale initiatives.

Example:

Tata Group:

Tata Group, one of India's oldest and most prestigious conglomerates, has a firm commitment to corporate social responsibility and sustainability. The organisation collaborates with multiple stakeholders to implement effective CSR programmes. The Tata Group collaborates with others for successful CSR planning as follows:

Tata Group interacts with the local communities in which its operations are located. The organisation conducts needs assessments and consultations to identify the most pressing community issues. Tata Group ensures that its CSR initiatives align with the specific requirements and priorities of the local population through this engagement.

 The Comprehensive Guide On CSR For Consultant

Partnerships with NGOs and Foundations: To implement CSR initiatives, Tata Group partners with reputable non-governmental organisations (NGOs) and foundations. These alliances utilise the expertise of non-governmental organisations (NGOs) in various social sectors, such as education, healthcare, and rural development, to develop effective and sustainable programmes.

Government Partnerships The Tata Group collaborates with government agencies to support national development priorities. By collaborating with the government, Tata Group's CSR initiatives can be incorporated into larger national development plans, ensuring improved coordination and long-term viability.

Tata Group supports social entrepreneurs and businesses that provide innovative solutions to societal issues. Through initiatives such as the Tata Social Enterprise Challenge, the organisation identifies and cultivates social entrepreneurs with scalable and sustainable business models for addressing social problems.

Tata Group contributes a substantial portion of its profits to charitable causes. The Tata Trusts, which were established by the founders of the Tata Group, play a vital role in philanthropic activities in a variety of fields, including healthcare, education, and poverty reduction.

Tata Group collaborates with international organisations and partners to resolve global issues such as climate change and environmental protection. These partnerships allow the company to align its CSR goals with global sustainability initiatives.

 The Comprehensive Guide On CSR For Consultant

Through volunteering and skill-based initiatives, the Tata Group involves its personnel in CSR activities. The company encourages its employees to support various social and environmental initiatives with their time and expertise.

Tata Group participates in multi-stakeholder platforms and industry associations in order to collaborate on CSR initiatives that have a collective effect on the industry and society.

Lastly, collaboration with multiple stakeholders enables businesses to develop CSR programmes that are more meaningful, sustainable, and beneficial to communities and the environment.

9.6: Develop a list of key performance indicators (KPIs)

Key Performance Indicators (KPIs) are quantifiable metrics that assist in evaluating the efficacy and impact of Corporate Social Responsibility (CSR) initiatives. Depending on the nature of the CSR programmes and the organization's objectives, the KPIs may vary. The following is a list of typical KPIs for CSR initiatives:

Number of Beneficiaries Reached	Determine the total number of people or communities who have directly benefited from CSR initiatives, such as education, healthcare, or livelihood programmes.
Percentage of Budget Spent on CSR	Track how much of the organization's profits or revenue goes to CSR efforts compared to the 2% that must be spent on CSR.
Employee	Find out how many hours workers

Volunteer Hours	spend doing volunteer work to show how committed the organisation is to helping the community.
Carbon Footprint Reduction	Find out how much greenhouse gas emissions have gone down because of sustainability projects, energy-saving steps, or the use of renewable energy.
Waste Reduction and Recycling	Track the percentage of waste diverted from landfills through recycling and waste reduction efforts
Social Return on Investment (SROI)	Evaluate the social and environmental impact of CSR initiatives in relation to the amount invested.
Stakeholder Engagement	Measure the level of engagement and satisfaction of stakeholders (employees, customers, communities, etc.) involved in or impacted by CSR activities.
Number of Partnerships with NGOs or Government Bodies	Evaluate the number of partnerships formed to expand the reach and efficacy of CSR initiatives.
Number of SDGs Addressed	Determine the extent to which CSR initiatives contribute to the achievement of specific SDGs outlined by the United Nations.
Diversity and Inclusion Metrics	Monitor the presence of different groups (gender, ethnicity, age, etc.) in the workforce and in positions of authority.
Health and Safety Incidents	Monitor and reduce the number of health and safety incidents among employees and communities impacted

 The Comprehensive Guide On CSR For Consultant

	by CSR activities.
Community Development	Measure improvements in community infrastructure, education, health, or other relevant parameters resulting from CSR projects.
Customer Perception and Loyalty	Survey customers to gauge their perception of the organization's CSR efforts and its impact on their loyalty and purchasing decisions.
Supplier Sustainability	Assess suppliers' adherence to supply chain sustainability policies and standards.
Media and Public Recognition	Track the extent to which the organization's CSR initiatives receive positive media coverage and public recognition.
Compliance and Adherence to CSR Regulations	Monitor the organization's compliance with CSR reporting requirements and adherence to relevant regulations.
Percentage of Projects Aligned with Core Business	Evaluate the extent to which CSR initiatives align with the organization's core business values and expertise.
Water Management and Conservation	Measure water usage reduction and water conservation efforts in the organization's operations.

Keep in mind that the selection of KPIs should be tailored to the particular objectives and context of each organization's CSR initiatives. Tracking and analysing these KPIs on a regular basis will provide valuable insights for enhancing the efficacy and impact of CSR programmes.

 The Comprehensive Guide On CSR For Consultant

Example

Creating a list of key performance indicators (KPIs) for effective corporate social responsibility (CSR) programmes entails creating particular metrics to monitor the impact and efficacy of CSR activities. Here's an example of a KPI list from a real Indian corporation, Hindustan Unilever Limited (HUL):

Hindustan Unilever Limited (HUL):

HUL is one of India's leading consumer products corporations, with a strong commitment to CSR and sustainability. The corporation prioritises CSR projects such as water conservation, women's empowerment, and health and hygiene programmes. Here's how HUL might develop a set of key performance metrics for its CSR initiatives:

Water Conservation:

KPI 1: Reduction in Water Consumption: Measure the percentage reduction in water consumption across manufacturing units and offices compared to the baseline year.

KPI 2: Water Recharge: Track the volume of rainwater harvested or groundwater recharged through water conservation projects.

Women Empowerment:

KPI 3: Women Workforce Representation: Measure the percentage increase in the representation of women in the workforce at all levels.

 The Comprehensive Guide On CSR For Consultant

KPI 4: Women in Leadership: Track the percentage of women in leadership positions within the company.

Health and Hygiene Programs:

KPI 5: Health Awareness: Measure the reach and impact of health awareness campaigns conducted by the company in partnership with local communities and NGOs.

KPI 6: Hygiene Infrastructure: Track the number of schools or community centers equipped with improved hygiene infrastructure, such as sanitation facilities.

Education and Skill Development:

KPI 7: Literacy Rates: Measure the improvement in literacy rates in communities where education initiatives are implemented.

KPI 8: Skill Training: Track the number of individuals trained and employed through skill development programs.

Environmental Impact:

KPI 9: Greenhouse Gas Emissions: Measure the reduction in greenhouse gas emissions per unit of production compared to the baseline year.

KPI 10: Waste Management: Track the percentage of waste recycled or diverted from landfills through waste management initiatives.

Community Engagement:

KPI 11: Community Outreach: Measure the number of community engagement events or programs conducted in a year.

 The Comprehensive Guide On CSR For Consultant

KPI 12: Community Satisfaction: Conduct surveys to assess the satisfaction level of communities with HUL's CSR efforts.

Supplier Sustainability:

KPI 13: Supplier Compliance: Assess the percentage of suppliers adhering to HUL's sustainability and ethical sourcing guidelines.

KPI 14: Supplier Capacity Building: Track the number of suppliers provided with training on sustainability practices.

Employee Engagement:

KPI 15: Employee Volunteering Hours: Measure the total number of volunteering hours contributed by HUL employees in CSR activities.

Tata Group:

One of India's largest and oldest companies, the Tata Group is noted for its dedication to social responsibility and sustainability. The Tata Group can create a list of key performance indicators (KPIs) for its CSR programmes as follows:

Education and Skill Development:

KPI 1: Number of Students Impacted: Measure the number of students benefiting from Tata's education initiatives, such as scholarships, vocational training, and digital learning programs.

KPI 2: Employability of Trained Candidates: Track the percentage of candidates trained by Tata Group's skill development programs who secure employment.

 The Comprehensive Guide On CSR For Consultant

Community Development:

KPI 3: Infrastructure Development: Measure the number of community projects undertaken, such as building schools, healthcare centers, or providing clean drinking water.

KPI 4: Livelihood Support: Track the number of livelihood support initiatives that help farmers and artisans in rural areas improve their income and livelihoods.

Renewable Energy and Environmental Sustainability:

KPI 5: Renewable Energy Capacity: Measure the installed capacity of renewable energy projects developed by Tata Group to reduce carbon emissions.

KPI 6: Waste Reduction: Track the percentage of waste recycled or upcycled through Tata's waste management initiatives.

Women Empowerment:

KPI 7: Women Workforce Representation: Measure the percentage of women in the workforce at Tata Group companies and their progression to leadership roles.

KPI 8: Women's Health and Hygiene: Track the reach and impact of health and hygiene programs specifically targeted towards women in rural and underserved areas.

Health and Sanitation:

KPI 9: Healthcare Accessibility: Measure the number of people benefiting from Tata's healthcare initiatives, including mobile health clinics and telemedicine services.

   The Comprehensive Guide On CSR For Consultant

KPI 10: Sanitation Coverage: Track the number of households gaining access to improved sanitation facilities through Tata's sanitation programs.

Rural Development:

KPI 11: Farmer Income Enhancement: Measure the percentage increase in farmers' income through agricultural training and market linkages provided by Tata Group.

KPI 12: Rural Electrification: Track the number of villages electrified through Tata's rural electrification projects.

Employee Volunteering and Engagement:

KPI 13: Employee Participation: Measure the percentage of employees actively participating in volunteering activities organized by Tata Group.

KPI 14: Volunteer Hours: Track the total number of hours contributed by employees in CSR-related volunteering.

Accessibility and Inclusivity:

KPI 15: Accessible Infrastructure: Measure the number of Tata Group facilities and workplaces made accessible to people with disabilities.

KPI 16: Diversity and Inclusion: Track the progress in promoting diversity and inclusion at all levels of the organization.

Tata Group & HUL can efficiently assess the effectiveness of its CSR programmes, align them with the company's values and long-term sustainability goals, and constantly improve its social and environmental contributions by establishing these KPIs.

 The Comprehensive Guide On CSR For Consultant

9.7: Develop a Project implementation Plan

Creating a Project Implementation Plan for corporate social responsibility (CSR) initiatives is crucial for a number of reasons.

For Corporate & Implementing Partner :

Project Overview and Objectives	Clearly state the goals of the CSR project, including its purpose, expected results, and how it fits into the company's CSR plan. Define the people and groups that the project is meant to benefit.
Stakeholder Analysis and Engagement	Identify essential stakeholders, such as employees, customers, local communities, non-governmental organisations, and government agencies. Develop a plan to engage stakeholders in the design and implementation of the project.
Resource Allocation	Allocate the necessary budget, manpower, and other resources required for the successful execution of the CSR project.
Project Timeline and Milestones	Develop a detailed timeline with specific milestones and deliverables for each phase of the project.
Risk Assessment and Mitigation	Identify hazards that could have an impact on the project's success and develop strategies to mitigate them.
Partnerships and Collaborations:	Explore the possibility of forming partnerships with NGOs, community organisations, and other stakeholders to increase the impact

   The Comprehensive Guide On CSR For Consultant

	of the initiative.
Project Activities and Implementation Strategy	Outline the specific steps that will be taken to accomplish the project's goals. Describe each activity's implementation strategy in detail.
Monitoring and Evaluation	Define key performance indicators (KPIs) to measure the progress and impact of the project. Create a monitoring and evaluation system Structure to routinely evaluate the project's effectiveness.
Communication and Reporting	Create a communication strategy to keep stakeholders apprised of the project's progress and results. Establish reporting mechanisms to provide management and stakeholders with periodic updates.
Capacity Building and Training	Identify any employee or community capacity-building requirements for the undertaking. Develop training programmes to improve the skills and knowledge necessary for the successful implementation of the project.
Sustainability and Exit Strategy	Create a plan to ensure the sustainability of the project beyond the initial implementation phase. Consider a project exit strategy that describes how the project's impact will be sustained or expanded after its completion.
Ethics and Compliance	Ensure that the project adheres to ethical practices, respecting the rights and dignity of beneficiaries

   The Comprehensive Guide On CSR For Consultant

	and stakeholders. Comply with all relevant laws, regulations, and reporting requirements related to CSR initiatives.
Continuous Improvement	Gather feedback from partners and use what you've learned in future projects to keep learning and getting better.
Final Evaluation and Impact Assessment	Conduct a final evaluation and impact assessment of the project to determine whether the objectives were achieved and the desired social impact was made.
Celebration and Recognition	Recognize and celebrate the achievements and success of the CSR project, acknowledging the efforts of all involved stakeholders.

By making a well-structured project implementation plan for CSR initiatives, companies can make sure that their social responsibility projects are carried out in a way that is effective and has a good effect on society.

Project Planning and Implementation

To show how project activities are carried out, let's think about a made-up social development project whose **goal is to give disadvantaged youth job training**. Here's an example of how the tasks of a project can be set up:

Activity 1: Needs Assessment and	**Task** • Conduct an in-depth need assessment to determine the

 The Comprehensive Guide On CSR For Consultant

Planning	specific skill deficiencies and training requirements of marginalised youth. • Engage with local community members, stakeholders, and potential beneficiaries to collect their input and insights. • Analyse the collected data and develop a detailed project plan based on the requirements and priorities identified. **Responsible : Project manager** **Duration: 1 month** **Required resources: include survey instruments, materials for community engagement, and data analysis software.**
Activity 2: Curriculum Development and Preparation of Training Materials	**Task** • Developing a vocational education programme in response to labour market needs and skill gaps is your assignment. • Create training courses, lessons, and training materials that aid in the acquisition of knowledge and abilities. • Include more real-world examples and interactive exercises in class to keep students interested and engaged. **Responsible: Curriculum Development Team** **Timeline: 2 months** **Resources Required: Subject matter experts, instructional**

	designers, training materials, multimedia resources.
Activity 3: Training Implementation	**Task** • Find appropriate locations with training equipment to hold the vocational classes. • To ensure the success of the training programme, it is necessary to find and train competent trainers or instructors to deliver it. • Plan and manage the logistics of the training sessions while keeping the attendees interested and involved. • Keep an eye on how things are going in the classroom and give your trainers the feedback they if need to improve. **Responsible: Training Coordinator** **Timeline: 6 months** **Resources Required: Training venues, training equipment, trainers, participant registration materials.**
Activity 4: Monitoring and Evaluation	**Task** • Develop a monitoring and evaluation framework to assess the effectiveness and impact of the vocational training program. • Collect data on participant attendance, satisfaction, skill development, and post-training

   The Comprehensive Guide On CSR For Consultant

	outcomes.
	• Analyze the collected data and measure the program's success against predefined key performance indicators (KPIs). • Use evaluation findings to identify areas for improvement and make necessary adjustments to the training program. **Responsible: Monitoring and Evaluation Team** **Timeline: Ongoing throughout the project duration** **Resources Required: Evaluation tools, data analysis software, monitoring forms.**
Activity 5: Graduation and Post-Training Support	**Task** • Organise a graduation ceremony to honour the trainees' accomplishments and distribute completion certificates. • Provide post-training support, such as employment placement assistance, entrepreneurship assistance, and mentoring programmes. • Create employment opportunities for the trained youth by forming partnerships with local businesses and organisations. **Responsible: Project Coordinator** **Timeline: Last month of the project**

 The Comprehensive Guide On CSR For Consultant

	Resources Required: Graduation ceremony arrangements, networking events, job placement resources.

These activities provide a general framework for implementing a vocational training-focused social development initiative. Depending on the nature of the undertaking, the available resources, and the needs of the intended beneficiaries, the particulars and timelines may vary. To guarantee the success and impact of the project, it is essential to adapt the activities to the project's context and continuously monitor and evaluate progress.

CHAPTER 10: METHODOLOGY FOR ORGANISING AND CARRYING OUT CSR PROJECTS

CSR projects need to be carefully planned, coordinated, and carried out to have the social and environmental effect that was intended. Here's a way to think about the process that might help:

10.1 Developing a CSR Strategy Based on circular economy

.**Principle of Circular Economy:** A circular economy is based on the idea that waste and pollution should be taken out of the system, goods and materials should be used more than once, natural systems should be restored, value should be kept and improved, collaboration and systems thinking should be encouraged, and renewable energy should be used. It aims to create a regenerative and sustainable economic model by getting rid of trash, encouraging reuse and recycling, restoring ecosystems, making the best use of resources, building partnerships, switching to renewable energy, and making sure everyone has the same rights. By following these principles, we can move away from the traditional linear "take-make-throw-away" model and towards a circular model that supports resource conservation, environmental stewardship, and long-term economic and social well-being.

Methodology to Developing a Circular Economy-Based CSR Strategy

Developing a CSR strategy based on the circular economy involves a thoughtful and systematic approach. Here are the steps to help you create an effective CSR strategy that aligns with circular economy principles:

Assess Current Practices	<ul><li>Examining past and present CSR initiatives of the company and their compatibility with Schedule VII of the Companies Act of 2013.</li><li>Analysing publicly accessible information regarding national and municipal development priorities in relation to CSR</li></ul>

	Sratergy.
	• Meeting with government and non-government development specialists to evaluate priorities and identify potential areas of involvement.
	• Evaluate your company's current waste generation, resource consumption, and product lifecycle practises. Determine where circular economy principles can be applied to enhance sustainability.
Set Clear Objectives	Define measurable and specific objectives for your CSR strategy. These could include reducing waste generation, increasing the proportion of recycled materials in products, and prolonging product life through repair and refurbishment.
Engage Stakeholders	Involve important stakeholders, including employees, customers, suppliers, investors, and local communities. Understand their sustainability and circular economy-related expectations and concerns. Their insights will be instrumental in formulating the strategy and garnering support.
Educate and Raise Awareness	Promote circular economy concepts within your organisation. Educate employees on the advantages of a circular economy and stress the importance of sustainability in decision-making.

Integrate Circular Design:	Encourage designers of products and packaging to begin with circular principles. Create items that are durable, repairable, and recyclable. Utilise renewable and repurposed materials whenever possible.
Implement measures to improve resource efficiency	Identify opportunities throughout your operations to increase resource utilisation and decrease waste production. Possibilities include improved inventory management, process optimisation, and waste reduction measures.
Collaborate and Form Partnerships	Collaborate with suppliers, clients, and other stakeholders to develop a circular value chain. Collaborate on recycling programmes, product take-back initiatives, and responsibility for circularity.
Circular activities Should Be Supported	Invest in or assist organisations and activities that advance the circular economy. This category could include partnerships with recycling facilities, investment in circular start-ups, and research on circular technology.
Measure and Monitor Progress	It is necessary to assess and monitor progress. Create KPIs to evaluate the effectiveness of your CSR strategy. Regularly monitor progress and disclose the outcomes both internally and externally. Transparent communication regarding your circular economy initiatives inspires stakeholder confidence.

 The Comprehensive Guide On CSR For Consultant

Innovate and Iterate	Create a culture of innovation and continuous improvement by innovating and iterating. Review and revise your CSR plan frequently in order to adapt to shifting circumstances, new technologies, and emergent best practises.
Celebrate Success and Share Learnings	Recognise accomplishments and share success stories with colleagues, customers, and the broader community. Demonstrating the beneficial results of your CSR strategy motivates others to follow suit.
Be Transparent and Communicate	Communicate openly about your CSR initiatives, progress, and issues. Engage stakeholders in discussions to offer updates and resolve any issues or feedback.

Keep in mind that developing a CSR strategy based on the circular economy is an exhausting and lengthy procedure. By incorporating circular economy principles into your business operations and engaging stakeholders, your organisation can make a significant contribution to sustainability while producing positive social and environmental impacts.

Example

Interface, Inc.

Interface is a leading manufacturer of commercial flooring. Historically, the corporation's manufacturing processes utilised vast quantities of nonrenewable resources and generated a substantial amount of waste.

CSR Strategy Based on Circular Economy:

Interface launched its ambitious Mission Zero programme in 1994, with the goal of eliminating the company's negative environmental influence by 2020. This strategy was completely consistent with the principles of the circular economy.

Interface is committed to procuring sustainable materials for use in its products. The company began manufacturing new carpet tiles from recycled materials, such as discarded fishing nets, old carpet tiles, and nylon refuse. This method decreased reliance on virgin resources and diverted waste away from landfills.

Interface introduced modular flooring products that could be installed, replaced, and reconfigured with ease. This strategy increased the flooring's lifespan, decreased the need for complete replacements, and enabled more efficient repair and refurbishment.

Product Take-Back Programme The company established a product take-back programme, "ReEntry," through which consumers could return used carpet tiles for recycling. Not only did this initiative enhance the recycling rate, but it also encouraged customers to participate in the circular economy.

Interface endeavoured to establish a closed-loop manufacturing procedure. This meant reprocessing old carpet tiles into new ones without sacrificing quality, thereby further reducing waste and consumption of resources.

 The Comprehensive Guide On CSR For Consultant

As part of its circular economy strategy, Interface has committed to becoming a carbon-neutral business. This entailed implementing energy efficiency measures, utilising renewable energy sources, and offsetting the remaining carbon emissions.

Interface collaborated with suppliers, clients, and industry partners to advance circular economy initiatives. The company actively pursued partnerships with recycling and sustainable materials-focused organisations.

Interface consistently reported on its progress towards attaining Mission Zero and other circular economy-related objectives. Transparent communication aided to build stakeholders' trust and credibility.

Result:

Interface accomplished its Mission Zero objective ahead of schedule in 2018, becoming carbon-neutral and eliminating negative environmental impacts.

The business reduced its carbon emissions by over 95%, its water consumption by 87%, and its refuse disposal by 92%.

Interface diverted significant waste from landfills by recycling over 340 million pounds of materials, including reclaimed fishing netting and post-consumer carpet.

Multiple times, Ethisphere Institute named the company one of the **"World's Most Ethical Companies"** in recognition of its efforts in the circular economy and sustainability.

Not only did Interface's CSR strategy founded on the circular economy help the company achieve impressive

 The Comprehensive Guide On CSR For Consultant

sustainability benchmarks, but it also bolstered the company's standing as an industry leader in responsible business practises. Interface is an outstanding illustration of how a real company can integrate circular economy principles into its CSR strategy to generate positive environmental and social outcomes.

Example 2:

Let's explore the CSR strategy of Patagonia, a well-known outdoor clothing and gear company that has been a pioneer in adopting circular economy principles.

Company: Patagonia

Patagonia is a renowned outdoor apparel company known for its commitment to environmental and social responsibility.

CSR Strategy Based on Circular Economy:

Worn Wear Program: Patagonia's "Worn Wear" programme encourages customers to purchase and sell used Patagonia clothing of high quality. By extending the lifecycle of its products and reducing the overall demand for new clothing, the company promoted a circular approach to consumption.

Repair and Recycle: The company actively encouraged consumers to repair rather than replace damaged Patagonia products. Patagonia offered free repair services for its apparel and equipment, encouraging customers to use their products for as long as feasible. Moreover, through its "Common Threads" recycling programme, Patagonia repurposed worn-out clothing into new products.

   The Comprehensive Guide On CSR For Consultant

Product Transparency: Patagonia prioritised transparency in its supply chain, informing consumers about the materials used in its products and the environmental and social impacts of their production. The company empowered consumers to make informed decisions and support sustainable practises by being transparent.

Sustainable Materials: Patagonia has invested in research and development to locate innovative and sustainable materials for its products. To lessen its reliance on nonrenewable resources, the company utilised recycled materials, organic cotton, and other eco-friendly alternatives.

Footprint Assessments: Patagonia regularly assessed the environmental footprint of its products and operations in order to identify areas for improvement. This strategy enabled the company to monitor its progress towards circular economy objectives and make decisions based on data.

Regenerative Organic Agriculture: Patagonia supported organic farming practises that prioritise soil health, biodiversity, and equitable treatment of farmers. The company utilised organic cotton in its products and promoted the adoption of regenerative business practises throughout its supply chain.

Advocacy and Activism: Patagonia advocated for environmental and social causes using its platform. The company actively participated in environmental campaigns, donated a portion of its sales to grassroots environmental organisations, and supported sustainability initiatives.

 The Comprehensive Guide On CSR For Consultant

B Corp Certification: Patagonia's certification as a B Corporation demonstrates its dedication to meeting stringent social and environmental performance standards. This certification demonstrated the organization's commitment to ethical business practises.

Results:

The "Worn Wear" programme effectively diverted a substantial quantity of clothing from landfills, thereby reducing overall waste and promoting a culture of reuse and repair.

Patagonia's initiatives to use recycled and sustainable materials reduced the company's environmental impact and served as a model for the apparel industry.

By prioritising transparency and advocacy, Patagonia developed a customer base whose values aligned with its own, resulting in increased brand loyalty and favourable word-of-mouth marketing.

The company's efforts in circular economy and sustainability earned it recognition as a leader in responsible business practises and a plethora of accolades.

Patagonia's CSR strategy founded on circular economy principles exemplifies how a business can embrace sustainable practises, engage with customers and stakeholders, and have a positive impact on the environment and society while still being commercially successful.

Example 3:

   The Comprehensive Guide On CSR For Consultant

Let's take a look at the CSR strategy of ITC Limited, an Indian conglomerate that has been actively incorporating circular economy principles into its business operations and sustainability initiatives.

Company: ITC Limited

ITC Limited is a diversified Indian conglomerate with interests in various sectors, including fast-moving consumer goods (FMCG), hotels, paperboards and packaging, agriculture, and information technology.

CSR Strategy Based on Circular Economy:

Sustainable Packaging: ITC has been actively working on sustainable packaging solutions for its FMCG products. The company has introduced innovative packaging materials, reduced single-use plastics, and increased the use of recycled and recyclable materials in its packaging.

Paper Recycling: As a major player in the paper and packaging industry, ITC has focused on promoting paper recycling. The company uses post-consumer recycled paper and promotes responsible paper usage among its stakeholders.

Solid Waste Management: ITC has implemented various solid waste management initiatives in its manufacturing facilities, hotels, and offices. The company emphasizes waste reduction, segregation, and recycling to minimize its environmental footprint.

Agri-Waste Utilization: In line with circular economy principles, ITC has been actively working on converting agri-waste into useful products. For example, the company

 The Comprehensive Guide On CSR For Consultant

has developed technologies to convert rice straw into pulp, which is then used in its paper manufacturing process.

Sustainable Agriculture: ITC has been promoting sustainable agricultural practices among farmers, encouraging responsible water usage, soil conservation, and integrated pest management techniques.

Social Forestry: The company has undertaken extensive social forestry projects to increase green cover, combat climate change, and provide sustainable livelihood opportunities to local communities.

Renewable Energy: ITC has invested in renewable energy sources, such as wind and solar power, to reduce its carbon footprint and transition towards a low-carbon future.

Eco-Friendly Hotels: ITC's hospitality division, ITC Hotels, has implemented sustainable practices in its properties, such as energy-efficient technologies, waste management systems, and responsible water usage.

Community Development: ITC's CSR initiatives also focus on community development, including education, healthcare, skill development, and women empowerment programs, which contribute to social well-being and inclusive growth.

Results:

ITC's sustainable packaging initiatives have led to a significant reduction in plastic usage and increased the use of recyclable materials in its product packaging.

The company's efforts in paper recycling and agri-waste utilization have contributed to a circular approach in its

 The Comprehensive Guide On CSR For Consultant

paperboards and packaging business, minimizing waste and resource consumption.

ITC's sustainable agricultural practices have positively impacted rural communities, enhancing farmers' livelihoods and promoting environmental conservation.

Through its renewable energy investments, ITC has reduced its carbon emissions, contributing to India's renewable energy goals.

The company's community development initiatives have had a positive social impact, supporting education, healthcare, and skill development in marginalized communities.

ITC Limited's CSR strategy based on circular economy principles demonstrates how a prominent Indian company can integrate sustainability into its core business activities while positively impacting the environment and society. By adopting circular economy practices and investing in responsible initiatives, ITC sets an example for other businesses to follow in India and beyond.

Nucor Corporation is an another example of a steel sector company that has adopted the circular economy approach. Nucor is one of North America's largest steel recyclers, and it has implemented a closed-loop system that reduces waste and maximises the use of recovered materials. When compared to typical blast furnaces, they have electric arc furnaces that can melt scrap steel, minimising the requirement for raw materials and energy. Nucor also works on circular product design, designing steel goods that can be reused and recycled. Nucor has become a leader in sustainable steel production as a result of these efforts.

   The Comprehensive Guide On CSR For Consultant

10.2: Operationalizing the institutional mechanism for sustainable Corporate social responsibility initiatives

The principle of operationalizing the institutional mechanism for sustainable Corporate Social Responsibility (CSR) initiatives includes committing to top-level management support, defining clear roles and responsibilities, establishing robust governance and policies, engaging stakeholders, integrating sustainability into the business strategy, measuring and reporting performance, and encouraging continuous improvement and innovation. By adhering to these principles, organisations can effectively integrate sustainability into their culture, strategy, and operations, thereby generating significant social and environmental impact and ensuring accountability and long-term sustainability.

> *Self-execution via an In-house CSR department: The Companies Act 2013, Section 135 and accompanying regulations permit the self-execution of CSR activities through an internal CSR department. Develop a CSR-specific department within your organisation. The department or foundation should consist of CSR, social impact evaluation, and community development specialists.*

> *Collaboration with non-governmental organizations: Making available funds to an independent implementation partner (with a three-year track record). Implementing CSR activities in collaboration with non-governmental organisations (NGOs) is a common and efficient method for corporate organisations in India to fulfil their CSR obligations under Section 135 of the Companies Act 2013.*

   The Comprehensive Guide On CSR For Consultant

Regardless of the structure chosen for implementation, the company must have a fundamental CSR department in place to support the CSR committee. The CSR committee will determine the function and structure of the department.

Selecting the Legal Structure of the Company Foundation In accordance with Section 135 of the Indian Companies Act of 2013, companies may establish a CSR Foundation as one of the lawful organisations to carry out their Corporate Social Responsibility (CSR) activities. Here is an overview of the CSR Foundation.

The CSR Foundation consists of: A CSR Foundation is a distinct legal entity established by a corporation for the sole purpose of conducting CSR operations. However, it is funded and maintained by the organisation. The primary objective of the CSR Foundation is to develop, implement, and monitor CSR initiatives in accordance with the company's CSR strategy and the 2013 Companies Act.

It allows funding from other sources, such as government initiatives and other foundations, to be leveraged. Typically, for-profit organisations are ineligible for such funding.Because the skills, job titles, career paths, and cost structures required for the implementation of CSR projects are substantially different from those required for a company's operations, a separate foundation permits the organisation to keep these distinct.

According to the proposed CSR regulations, an organisation established by a corporation to assist in the implementation of its CSR operations must be registered in India under section 818 of the Companies Act, 2013 as a trust, society, or non-profit company. In India, non-

 The Comprehensive Guide On CSR For Consultant

Establishing the Legal Structure

Establishing a legal entity in India under the Trust, Society, or Section 8 (Non-Profit) Company Act to fulfil corporate social responsibility (CSR) requirements entails numerous stages. The following is a step-by-step guide to help you through the process:

Selecting the Legal Structure: Choose the appropriate legal structure based on your CSR objectives and long-term goals. The options include a Trust, Society, or Section 8 (Non-Profit) Company.

Consider factors such as ease of registration, governance structure, compliance requirements, and tax implications before finalizing the legal entity.

Drafting the Memorandum of Association (MOA, AOA) and Rules/Bylaws: Prepare the MOA and rules/bylaws (depending on the chosen legal structure) outlining the purpose, objectives, and operational Structure of the entity.

Ensure that the CSR policy is integrated into the MOA/rules/bylaws to provide a clear Structure for CSR activities.

Registration and Legal Compliance: Register the entity with the appropriate regulatory authority (e.g., Registrar of

 The Comprehensive Guide On CSR For Consultant

Trusts, Registrar of Societies, or Registrar of Companies for Section 8 companies).

Comply with all the legal requirements, documentation, and filings needed for registration.

Accounting and Finance Systems: Implement robust accounting systems to track and manage the financial transactions related to CSR activities separately from regular business activities.

Ensure proper bookkeeping and financial reporting, including a separate balance sheet and income statement for CSR expenses.

Tax Exemption and Compliance: Apply for tax exemption status under Section 12A and 80G of the Income Tax Act to avail tax benefits for the entity and donors supporting CSR initiatives.

Ensure compliance with the tax laws and timely filing of tax returns.

Administration and HR Alignment: Establish a clear organizational structure, including governance bodies and roles.

Define the roles and responsibilities of staff involved in managing and implementing CSR initiatives.

Align HR policies to recruit and retain employees committed to the organization's CSR goals.

M&E Systems Integration: Set up M & E systems to monitor and evaluate the impact of CSR projects and track the utilization of funds.

 The Comprehensive Guide On CSR For Consultant

Use technology for efficient communication and collaboration among team members.

Stakeholder Engagement: Engage with stakeholders, including communities, government agencies, NGOs, donors, and volunteers, to build strong partnerships and support for CSR projects.

Measuring and Reporting: Establish a robust monitoring and evaluation Structure to measure the outcomes and impact of CSR initiatives.

Regularly report on CSR activities to stakeholders and the public through annual reports and other communication channels.

Continuous Improvement: Continuously assess the effectiveness of CSR projects and make improvements based on feedback and learnings.

Legal and Regulatory Compliance: Stay updated with changes in relevant laws and regulations related to CSR activities and ensure compliance.

Creating a legal corporation for CSR projects involves time, effort, and adherence to legal and financial standards. You may effectively deliver on the commitments set in the CSR policy and generate a positive and sustainable influence on society and the environment by harmonising the various systems and roles.

Example

Adani Foundation

Establishing the Adani Foundation has helped the Adani Group implement corporate social responsibility (CSR) initiatives in numerous ways:

The Adani Foundation is a centralised organisation responsible for organising, implementing, and monitoring CSR initiatives across all Adani Group companies. This centralised strategy assures improved CSR coordination and alignment.

Focused Areas: The Adani Foundation can focus on specific focal areas that align with the CSR policy and priorities of the company. This targeted approach enables for a greater depth and significance of impact in specific social and environmental areas.

Community Engagement: The Foundation facilitates direct engagement with local communities to better comprehend their requirements and aspirations. This ensures that CSR projects are pertinent and meet the needs of the community.

Infrastructure and Resources: The Adani Foundation provides the infrastructure, resources, and expertise required to efficiently execute CSR initiatives. This consists of financial resources, qualified personnel, and access to technology and knowledge.

The Foundation can collaborate with government bodies, non-profit organisations, and other stakeholders to leverage their expertise and resources. Such partnerships increase the scope and efficacy of CSR initiatives.

The establishment of the Foundation demonstrates the Adani Group's commitment to social and environmental causes over the long term. This commitment promotes credibility and trustworthiness among stakeholders, such as communities, investors, and regulators.

Transparency and Accountability: The Adani Foundation's CSR activities are transparent. Reporting, audits, and public disclosures on a regular basis demonstrate accountability and compliance with regulatory requirements.

Employee Engagement: Through volunteer programmes and skill-based initiatives, the Foundation encourages employee engagement in CSR initiatives. Engaging employees in social initiatives enhances their sense of purpose and cultivates a culture of giving back.

The Foundation conducts impact assessments to evaluate the efficacy and results of CSR initiatives. This aids in refining strategies and assuring efficient resource utilisation.

Recognition and Brand Reputation: The Adani Group's CSR initiatives enhance its reputation as a socially responsible business. The Foundation's activity contributes to the development of a positive brand image and fosters community goodwill.

The Foundation encourages innovation in addressing social and environmental challenges. Successful pilot initiatives can be expanded across multiple regions, thereby maximising their global impact.

In accordance with the Indian Companies Act, the Adani Foundation ensures compliance with CSR-related laws and guidelines. This includes spending the required amount on CSR activities and reporting in accordance with regulatory guidelines.

   The Comprehensive Guide On CSR For Consultant

Overall, the Adani Foundation plays a crucial role in directing the Adani Group's resources and efforts towards making a positive impact in the communities where they operate and contributing to long-term development. It enables the organisation to align its business goals with social and environmental objectives, resulting in a more responsible and inclusive business culture.

10.3: Due-Diligence of the Implementation Partner

The principle of undertaking due diligence on the implementation partner for Corporate Social Responsibility (CSR) initiatives requires a comprehensive evaluation of the partner's suitability and alignment with the organization's CSR objectives. This includes assessing their track record, financial stability, and legal compliance, as well as assuring transparency, accountability, and ethical conduct. It is essential to evaluate the partner's expertise, capacity, and experience in implementing CSR initiatives, taking into account their technical expertise and resources. Engaging stakeholders and collaborating with pertinent parties are crucial, as they ensure the partner's ability to establish positive relationships and manage potential conflicts of interest. In addition, verifying the partner's risk management practises and regulatory compliance is essential for mitigating the initiatives' potential risks. Lastly, evaluating their measurement and reporting capabilities ensures accurate reporting on the outcomes and efficacy of CSR initiatives, thereby providing stakeholders with transparency and accountability.

By adhering to these principles, organisations can select an appropriate implementation partner for their CSR initiatives, one that is aligned with their values, capable of

 The Comprehensive Guide On CSR For Consultant

executing the initiatives effectively, and committed to delivering positive social and environmental impacts while adhering to ethical and responsible practises.

Legal Compliance	The process of due diligence helps the corporation in ensuring that the partner or NGO is registered and in compliance with all applicable regulations governing NGOs and charity organisations in India.
Alignment with CSR Goals	Due diligence is used to find out if a partner's purpose, vision, and past projects are in line with the company's CSR goals and areas of focus. This makes sure that the partnership will have a big and important effect on society.
Impact Assessment Capability	A comprehensive method of due diligence evaluates the partner's capacity to assess and report the impact of their social projects. This skill is essential for comprehending the effectiveness and outcomes of CSR initiatives.
Operational Capacity	Due diligence assists in evaluating the operational capacity of the partner, including their team's expertise, project management skills, and implementation experience. This evaluation ensures that the proposed CSR projects can be executed effectively by the partner.
Transparency and Accountability	The due diligence process seeks to ascertain the partner's level of transparency and accountability in

 The Comprehensive Guide On CSR For Consultant

	financial reporting, project execution, and utilization of resources.
Risk Mitigation	Due diligence helps the company figure out what risks might be involved in the partnership. Taking care of these risks right away helps reduce bad outcomes and possible damage to Company brand.
Stakeholder Engagement	Evaluating the partner's stakeholder involvement practises makes sure that their projects are focused on the community and that the people who benefit from them are involved in decision-making.
Legal and Ethical Compliance	The due diligence process checks if the partner adheres to ethical practices and does not engage in activities that may conflict with the company's values or legal requirements.

Overall, it is important to do due diligence on the CSR implementation partner or NGO to make sure that the CSR funds and efforts are used in a responsible and effective way. It helps build a partnership that is good for both sides and has an effect, which is good for India's long-term social development.

Let's look at a real company as an example of how to do due research on a partner or NGO that is helping to implement a company's CSR plans.

Example:

Hindustan Unilever Limited (HUL)

 The Comprehensive Guide On CSR For Consultant

Background: Hindustan Unilever Limited (HUL) is one of India's largest consumer goods companies, offering a wide range of products across various categories. HUL has a strong commitment to CSR and sustainability, with initiatives focused on health and hygiene, water conservation, education, women empowerment, and rural development.

Steps in Conducting Due Diligence:

Identifying Potential Partners: HUL begins by identifying potential NGO partners that align with its CSR goals and focus areas. The company looks for organizations that have a proven track record of impactful work and share similar values.

Reviewing Legal and Regulatory Compliance: HUL ensures that the selected NGOs are registered under the appropriate laws and regulatory bodies in India. They verify the validity of their registration and compliance with all relevant laws governing NGOs.

Assessing Past Projects and Impact: HUL reviews the past projects executed by the potential NGOs to understand their scope, effectiveness, and impact. They analyze the data and outcomes to assess the organizations' ability to create sustainable social change.

Financial Evaluation: HUL conducts a financial evaluation of the NGOs to ensure their financial stability and capacity to manage CSR funds. This includes reviewing audited financial statements and understanding the organization's sources of funding.

 The Comprehensive Guide On CSR For Consultant

Governance and Transparency: HUL examines the governance structure of the NGOs, including their board members and leadership. They look for transparency in financial reporting and adherence to ethical practices.

Operational Capacity and Expertise: HUL evaluates the NGO's operational capacity, including the qualifications and expertise of their team members. They assess whether the organization has the necessary resources and skills to implement the proposed CSR initiatives effectively.

Stakeholder Engagement: HUL assesses the NGO's stakeholder engagement practices to ensure that they involve and empower local communities in their projects. This involvement is critical for the sustainability and relevance of the initiatives.

Example of NGO Partner Selected by HUL:

Let's assume that HUL identifies an NGO called **"Health for All"** that focuses on providing healthcare services to underserved communities in rural areas.

Reasons for Selection:

"Health for All" is a registered NGO under the Indian laws and complies with all regulatory requirements.

The NGO has successfully implemented healthcare projects in several villages, significantly improving the health and well-being of the communities.

Their financial statements show stable financial management, with diverse funding sources, including grants and individual donations.

   The Comprehensive Guide On CSR For Consultant

The organization has a transparent governance structure, and their projects involve active participation from the local community members.

"Health for All" has a dedicated team of healthcare professionals and volunteers who are experienced in managing healthcare initiatives in rural settings.

Benefits of Conducting Due Diligence:

Due diligence on "Health for All" guarantees that HUL collaborates with a competent and capable NGO capable of effectively implementing healthcare programmes aligned with HUL's CSR goals. The due diligence approach reduces risks and increases the likelihood of having a good and long-term influence on the targeted communities. Furthermore, it gives transparency and responsibility to HUL's stakeholders while demonstrating ethical use of CSR funding.

Due diligence procedures for selecting an international implementation partner (Only for Reference)

There are a number of concepts and regulations in the international standards for doing partner agency due diligence, especially in the context of corporate social responsibility (CSR) programmes. The following are some of the most important global norms and structures:

UN Guiding Principles on Business and Human Rights (UNGP): The UNGP provides a global standard for preventing and addressing the adverse human rights impacts of business activities. It emphasizes the responsibility of companies to conduct due diligence to

   The Comprehensive Guide On CSR For Consultant

identify, prevent, mitigate, and account for how they address human rights risks.

ISO 26000 - Social Responsibility: ISO 26000 is an international standard that provides guidance on social responsibility. It emphasizes the importance of stakeholder engagement and due diligence in understanding the social, environmental, and ethical impacts of business activities.

OECD Guidelines for Multinational Enterprises: The OECD Guidelines offer recommendations to multinational enterprises for responsible business conduct. The guidelines emphasize due diligence as a key process to identify and address potential adverse impacts on society and the environment.

International Finance Corporation (IFC) Performance Standards: The IFC Performance Standards are a set of social and environmental requirements for private sector projects. They include specific guidance on conducting due diligence to assess and manage risks and impacts.

Global Reporting Initiative (GRI) Standards: The GRI Standards provide guidelines for sustainability reporting. It encourages organizations to disclose information about their due diligence processes related to social and environmental impacts.

ISO 37001 - Anti-Bribery Management Systems: While not directly related to CSR, ISO 37001 focuses on preventing bribery and promoting an ethical culture. It encourages due diligence to identify bribery risks and implement appropriate measures.

 The Comprehensive Guide On CSR For Consultant

ILO Tripartite Declaration of Principles concerning Multinational Enterprises and Social Policy: This declaration outlines principles related to labor rights, social protection, and industrial relations. It emphasizes the importance of due diligence in ensuring respect for labor standards.

Equator Principles: These are a set of environmental and social risk management guidelines for financial institutions for assessing and managing the environmental and social risks of project financing.

The United Nations Global Compact (UNGC): While not a specific standard, the UNGC provides a Structure for businesses to commit to sustainable and socially responsible practices, including due diligence processes.

These global guidelines emphasise the importance of conducting due diligence as part of responsible business practises, especially when working with government agencies on CSR initiatives. The process of due diligence aides in the identification and mitigation of potential risks and consequences, promotes transparency, and contributes to long-term growth and social welfare. Companies are strongly encouraged to utilise these guidelines and adapt their due diligence procedures to the specific context and challenges of CSR initiatives.

10.4: Corporate Social Responsibility Project Baseline study

Corporate Social Responsibility (CSR) project baseline study involves establishing a solid foundation and understanding of the current social, environmental, and economic conditions before implementing CSR initiatives.

 The Comprehensive Guide On CSR For Consultant

This study serves as a reference point against which the project's impact and progress can be measured. The baseline study should encompass the following principles:

Comprehensive Assessment: Conduct a full evaluation of the project area, taking into account social, economic, and environmental factors. This means gathering data and information on key factors like poverty rates, levels of education, environmental degradation, community needs, and the infrastructure that is already there. For the assessment, local stakeholders like community members, NGOs, and government bodies should be consulted.

Longitudinal Perspective: Take a long-term perspective of the project area by keeping track of its history and trends. This shows how the problems, changes, and success have altered over time. By knowing how things were before and what happened in the past, it's possible to figure out what's really going on and set the right objectives and targets for CSR projects. The study should include both qualitative and quantitative data, such as surveys, interviews, focus group talks, and data from secondary sources.By adhering to these principles, organizations can conduct a robust

baseline study for CSR projects. This study will provide a clear understanding of the project's starting point, inform goal setting and strategy development, and enable the monitoring and evaluation of the project's impact over time. It also ensures that the CSR initiatives are designed and implemented based on the specific needs and context of the project area, fostering meaningful and sustainable outcomes.

Process of Baseline study

 The Comprehensive Guide On CSR For Consultant

Conducting a baseline study is essential prior to developing a CSR (Corporate Social Responsibility) project because it provides essential data and information on the Current Situation, Setting Realistic Targets and Objectives, Evaluating Project Impact, Resource Allocation, Community Stakeholder Engagement, Identifying Key Indicators, Risk Assessment, and Project Design and Tailoring that serves as the foundation for project planning, implementing, and evaluating the project's impact.

Define the objectives	Clearly outline the objectives of the CSR initiatives and the specific outcomes you want to measure through the baseline survey. Determine the key indicators that will help you assess the impact of the initiatives effectively.
Identify the target population	Define the target population or beneficiaries of the CSR initiatives. This could include employees, local communities, specific demographic groups, or other stakeholders who will be directly affected by the initiatives.
Develop the survey instrument	Create a structured survey questionnaire that aligns with the objectives and indicators of the CSR initiatives. The survey should include both quantitative and qualitative questions to gather comprehensive data.
Pre-test the survey	Before implementing the baseline survey, pre-test the questionnaire with a small sample of the target population to identify any issues with

     The Comprehensive Guide On CSR For Consultant

	the questions' clarity, relevance, or structure. Make necessary adjustments based on the feedback received.
Obtain ethical clearance	If the survey involves sensitive data or human subjects, ensure you obtain ethical clearance from relevant authorities.
Train surveyors	If you are using a team of surveyors to collect data, provide them with comprehensive training on the survey instrument, data collection protocols, and ethical considerations.
Data collection	Administer the survey to the target population using appropriate data collection methods, such as face-to-face interviews, group discussions, phone surveys, online surveys, or a combination of these methods. Ensure confidentiality and anonymity of respondents if necessary.
Triangulation of data	Use multiple sources of data, such as official records, existing reports, and secondary data, to complement the survey findings and ensure the reliability of the baseline data.
Data analysis	Process and analyze the collected data using statistical software or qualitative analysis techniques, depending on the nature of the data. Calculate descriptive statistics and identify key patterns and trends.
Create a baseline report	Prepare a comprehensive baseline report that includes the methodology, data collection process, key findings,

   The Comprehensive Guide On CSR For Consultant

	and limitations of the survey. The report should serve as a reference point for measuring the impact of the CSR initiatives in the future.
Ensure data integrity	Validate the accuracy and integrity of the data by cross-checking and verifying the information collected during the survey.
Use findings for program design	Utilize the baseline survey findings to inform the design and implementation of the CSR initiatives. Tailor the programs based on the specific needs and challenges identified in the baseline data.

Remember that the baseline survey serves as the basis for evaluating the effectiveness of CSR initiatives. By conducting a comprehensive and well-designed baseline survey, businesses can gain a better understanding of the current situation, establish realistic goals, and make decisions based on data to create positive social and environmental impacts.

Tools and Methodologies to gather and analyze relevant data

Conducting a baseline study for CSR project development requires the use of various tools and methodologies to gather and analyze relevant data. The choice of tools depends on the project's scope, objectives, and the specific data needed. Here are some common tools and techniques used for conducting baseline studies in CSR projects:

    The Comprehensive Guide On CSR For Consultant

Surveys and Questionnaires: Surveys are effective for gathering quantitative and qualitative data from a large number of respondents. They can be administered through face-to-face interviews, phone calls, or online platforms. Questionnaires help capture baseline information on demographics, socio-economic conditions, and perceptions of beneficiaries and stakeholders.

Interviews: Conducting in-depth interviews with key informants, community leaders, and other stakeholders provides valuable qualitative insights into the local context, needs, and challenges.

Focus Group Discussions (FGDs): FGDs involve small group discussions with community members to explore their perceptions, experiences, and opinions related to the CSR project's focus areas.

Participatory Rural Appraisal (PRA) Techniques: PRA techniques, such as social mapping, seasonal calendars, and resource mapping, involve engaging with community members to collectively gather information and insights about their environment and needs.

Key Informant Interviews (KII): KIIs involve interviewing key stakeholders, experts, and local authorities to gain an in-depth understanding of the project's context and potential challenges.

Secondary Data Review: Analyzing existing data, such as census data, government reports, and academic studies, provides background information on the project area and assists in identifying trends and patterns.

 The Comprehensive Guide On CSR For Consultant

Observations: Direct observation of the project area and community activities can offer valuable contextual information and identify potential project opportunities and challenges.

Geographic Information Systems (GIS): GIS tools can help map and analyze spatial data, which is useful for understanding geographical patterns and potential project impact areas.

Baseline Surveys and Assessments: Specific assessment tools are available for different development sectors, such as education, health, environment, and livelihoods. These tools are designed to measure baseline indicators and track changes over time.

Participatory Impact Assessments: Participatory tools, such as outcome mapping and Most Significant Change (MSC) techniques, involve beneficiaries and stakeholders in defining project goals and assessing outcomes.

Social Impact Assessments (SIAs): SIAs are used for larger CSR projects to assess their potential social impacts, both positive and negative, and develop mitigation strategies.

When conducting a baseline study, it is essential to use a combination of these tools to ensure a comprehensive and accurate understanding of the project's context and needs. Additionally, involving local communities and stakeholders in the data collection process fosters inclusivity and ensures that the project's design is aligned with their needs and priorities.

10.5: Corporate Social Responsibility Project Design & Development

 The Comprehensive Guide On CSR For Consultant

Developing the design for a CSR (Corporate Social Responsibility) project involves several tools and methodologies to ensure a comprehensive and well-structured approach. Here are some essential tools commonly used in CSR project design:

Needs Assessment and Stakeholder Analysis:

Identify the social and environmental issues within the target community or region where the CSR project will be implemented. Conduct a comprehensive needs assessment to understand the specific challenges and priorities of the community. Additionally, perform a stakeholder analysis to identify key stakeholders, their interests, and their level of influence in the project.

Goal and Objectives:

Define the overarching goal of the CSR project, which should align with the organization's CSR strategy and address the identified needs. Develop clear and measurable objectives that contribute to achieving the goal. Objectives should be specific, attainable, relevant, and time-bound (SMART).

Theory of Change or Logic Model:

Create a Theory of Change (ToC) or logic model that outlines the causal pathways through which the CSR project activities will lead to desired outcomes and impacts. This visual representation helps in understanding the project's strategy and aligning activities with intended results.

Project Activities and Interventions:

 The Comprehensive Guide On CSR For Consultant

Design the specific activities and interventions that will be implemented to achieve the project objectives. Ensure that these activities are well-suited to address the identified needs and are aligned with the organization's expertise and resources.

Budget and Resource Planning:

Develop a detailed budget for the CSR project, taking into account the costs of activities, personnel, materials, and any external resources required. Allocate resources effectively to ensure the successful implementation of the project.

Partnerships and Collaborations:

Identify potential partners and collaborators, such as local NGOs, government agencies, and other stakeholders. Foster strong partnerships to leverage expertise, resources, and local knowledge for more significant impact.

Monitoring and Evaluation (M&E) Structure:

Establish a robust M&E Structure to track the progress and impact of the CSR project. Define indicators, data collection methods, and evaluation timelines. Regularly monitor the project's implementation to identify any challenges and make necessary adjustments.

Risk Assessment and Mitigation:

Conduct a risk assessment to identify potential challenges and risks that may impact the project's success. Develop mitigation strategies to address these risks and ensure project continuity.

Community Engagement and Participation:

Involve the target community and beneficiaries in the project design and development process. Engage them in decision-making, planning, and implementation to ensure project ownership and sustainability.

Communication and Reporting:

Develop a communication plan to keep stakeholders informed about the progress and impact of the CSR project. Regularly report on project achievements, challenges, and outcomes to internal and external stakeholders.

Ethical Considerations:

Ensure that the CSR project adheres to ethical principles and respects the cultural, social, and environmental norms of the target community. Implement responsible business practices throughout the project lifecycle.

Sustainability and Exit Strategy:

Integrate sustainability principles into the project design to ensure its long-term impact. Develop an exit strategy that outlines how the project will be handed over to the community or other stakeholders after its completion.

Legal and Regulatory Compliance:

Ensure compliance with all relevant laws, regulations, and guidelines related to CSR activities in the project's implementation location.

Continuous Learning and Adaptation:

Emphasize a culture of continuous learning and adaptation throughout the project. Use feedback from beneficiaries

   The Comprehensive Guide On CSR For Consultant

and stakeholders to improve project strategies and outcomes.

Impact Assessment:

Plan for impact assessment and evaluation at various stages of the project to measure its effectiveness and success in achieving the intended objectives and impacts.

In conclusion, a well-structured Structure for CSR project design and development is crucial for ensuring that CSR initiatives are aligned with the organization's goals, address community needs, and achieve sustainable and impactful outcomes. The Structure should be flexible enough to adapt to changing circumstances and promote continuous improvement in CSR practices.

10.6: Finalizing the Arrangement with the Implementing Agency

Finalizing the arrangement with the CSR initiative implementing agency is a critical step to ensure the successful execution of the CSR project. Here are the key steps to finalize the arrangement with the implementing agency:

Example: Finalizing the Arrangement with the CSR Initiative Implementing Agency - "Project Bright Future"

Conduct a thorough evaluation: Before finalizing the agency, conduct a comprehensive evaluation of potential implementing partners. Assess their track record, experience, expertise, and alignment with the goals and values of the CSR initiative.

 The Comprehensive Guide On CSR For Consultant

Example: Company XYZ, a multinational corporation, is committed to improving access to education for underprivileged children. They have identified "Project Bright Future" as their CSR initiative to build and equip a learning center in a low-income community. After conducting a rigorous evaluation of potential implementing agencies, they have selected "Education for All Foundation" (EFAF) as their partner. EFAF has a proven track record of successfully implementing similar educational projects in disadvantaged communities.

Define roles and responsibilities: Clearly outline the roles and responsibilities of both parties in a written agreement. Define the scope of work, project timelines, deliverables, and reporting requirements.

Example: Company XYZ and EFAF work together to clearly define their roles and responsibilities. Company XYZ will provide the funding and resources required for the construction of the learning center, while EFAF will be responsible for the overall implementation, including hiring and training teachers, curriculum development, and monitoring the project's impact.

Set performance indicators: Establish measurable performance indicators and targets that the implementing agency will be accountable for. This will ensure transparency and facilitate monitoring and evaluation of the project's progress.

Example: Key performance indicators (KPIs) are established to measure the success of "Project Bright Future." These KPIs include the number of children enrolled, attendance rates, academic performance, and

   The Comprehensive Guide On CSR For Consultant

community engagement. EFAF commits to regular reporting on these metrics to Company XYZ.

Establish a budget and funding agreement: Agree on the project budget, funding sources, and payment terms. Ensure that the agency has a clear understanding of financial reporting requirements and adheres to proper financial management practices.

Example: Company XYZ and EFAF collaboratively finalize the budget for "Project Bright Future," considering construction costs, operational expenses, and potential contingencies. The funding agreement outlines the disbursement schedule and payment terms, ensuring transparency in financial matters.

Develop a communication plan: Agree on a communication plan that outlines regular updates, progress reports, and channels of communication between the company and the implementing agency.

Example:

Both parties agree on a communication plan that includes regular progress updates, quarterly meetings, and an annual report showcasing the project's impact. They also decide to hold community meetings to involve local stakeholders and ensure their input is considered.

Ensure compliance and legal aspects: Ensure that the implementing agency complies with all legal and regulatory requirements related to the CSR project. Establish necessary contracts and agreements to safeguard both parties' interests.

 The Comprehensive Guide On CSR For Consultant

Example: Company XYZ ensures that EFAF meets all legal and regulatory requirements for implementing the CSR initiative. Both parties sign a formal agreement outlining their responsibilities, obligations, and expectations.

Build a relationship of trust: Foster a relationship of trust and collaboration with the implementing agency. Regularly engage in dialogue, provide support, and address any challenges that arise during the project implementation.

Example: Company XYZ and EFAF foster a relationship of trust and collaboration. They establish open channels of communication and conduct regular site visits to monitor the progress of the learning center's construction and program implementation.

Monitor and evaluate: Establish a monitoring and evaluation mechanism to assess the project's progress and impact. Regularly review the performance of the implementing agency and provide constructive feedback.

Example: EFAF implements a robust monitoring and evaluation system to track the project's progress against the predefined KPIs. Regular reviews with Company XYZ help identify any challenges and take corrective actions promptly.

Support capacity-building efforts: Offer capacity-building support to the implementing agency if needed. This may include training sessions, workshops, or access to additional resources to enhance their effectiveness.

Example: Company XYZ offers support to EFAF in the form of additional training for teachers and staff, as well as access to learning resources and educational materials.

   The Comprehensive Guide On CSR For Consultant

Review and adapt: Continuously review the progress of the CSR initiative and be open to making necessary adjustments. Be willing to adapt the approach based on lessons learned and changing circumstances.

Example: Both parties continuously review the project's progress and remain open to making necessary adjustments. They collaborate to address any unforeseen challenges and adapt the project strategy as needed.

Celebrate successes: Recognize and celebrate the achievements of the implementing agency and their contributions to the CSR initiative. Acknowledge their efforts and commitment to the project.

Example: Company XYZ publicly acknowledges the successes of "Project Bright Future" through press releases, social media posts, and community events. They also extend appreciation to EFAF and the local community for their contributions to the initiative's success.

Foster long-term partnerships: Consider establishing a long-term partnership with the implementing agency if the CSR initiative's goals align with ongoing social development efforts. A sustainable partnership can lead to greater impact and continuity of positive outcomes.

Example: Impressed with EFAF's dedication and impact, Company XYZ decides to establish a long-term partnership. They commit to supporting EFAF's efforts in the community beyond "Project Bright Future" and explore the potential for joint projects in the future.

By following these steps and maintaining effective communication and collaboration with the implementing

 The Comprehensive Guide On CSR For Consultant

agency, the company can ensure a successful and impactful CSR initiative that creates positive change in the community or environment it aims to serve.

10.7: Corporate Social Responsibility Project Proposal Approval Process

The approval procedure for an NGO project proposal for a Corporate Social Responsibility (CSR) initiative consists of multiple steps that the CSR committee uses to evaluate the proposal's alignment with the company's CSR strategy and its potential social impact. The typical phases of the approval procedure are listed below:

Submission of Proposal	The NGO submits a comprehensive project proposal to the CSR department of the company. The proposal should describe the project's objectives, activities, intended recipients, budget, schedule, and anticipated outcomes.
Preliminary Review	The CSR department conducts an initial review of the proposal to ensure that it satisfies the basic eligibility requirements and aligns with the organization's CSR policy and priorities.
Internal Evaluation	The CSR committee thoroughly evaluates the proposal. They evaluate its relevance to the company's CSR objectives, its potential impact on the community, and its implementability.
Stakeholder Consultation	The CSR committee may engage in discussions with the NGO and

 The Comprehensive Guide On CSR For Consultant

	conduct site visits in order to comprehend the local context, the beneficiaries' requirements, and the NGO's capabilities.
Due Diligence	The CSR committee could perform due diligence on the NGO in order to evaluate its credibility, track record, financial stability, and governance practises.
Budget Allocation	Once the CSR committee is satisfied with the proposal, they determine the budget allocation for the project based on the estimated cost and the available CSR funds.
Board Approval	For significant CSR initiatives, the Board of Directors may need to approve the project proposal and budget allocation. The proposal is presented to the board for evaluation and approval by the CSR committee.
Agreement and MoU	The company and the NGO formalise the collaboration with a Memorandum of Understanding (MoU) or agreement following approval. The MoU defines each party's roles, responsibilities, and expectations.
Project Implementation	With the approval and agreement in place, the NGO proceeds with the implementation of the project as per the agreed-upon plan and timeline.
Monitoring and Evaluation	Throughout the project implementation, the CSR committee monitors the progress and impact of

   The Comprehensive Guide On CSR For Consultant

	the project. The NGO provides regular updates on the project's achievements and challenges.
Reporting and Communication	The NGO submits periodic progress reports to the CSR committee, updating them on the project's activities, expenditure, and outcomes. The committee communicates these updates to relevant stakeholders, including senior management and the Board.
Impact Assessment	At the end of the project or during its course, the CSR committee may conduct an impact assessment to evaluate the project's success in achieving its intended outcomes.
Closure and Documentation	The NGO submits a final report detailing the project's impact, lessons learned, and recommendations following its conclusion. This documentation contributes to the organization's CSR reporting and helps inform future CSR initiatives.

The approval procedure ensures that the selected NGO project is consistent with the corporation's CSR strategy, complies with applicable regulations, and has the potential to have a positive and lasting impact on society.

10.8: Employee Engagement in Corporate Social Responsibility Initiatives

Enhancing employee participation in corporate social responsibility (CSR) initiatives is crucial for fostering a culture of social responsibility and sustainability within an organisation. When employees are actively involved in CSR efforts, they experience a greater sense of purpose, greater pride in their work, and a stronger connection to the organization's mission. Here are a few strategies for increasing employee participation in CSR initiatives:

Employee Engagement IN CSR

Align CSR with Company Values: Ensure that the CSR initiatives align with the company's fundamental values and overall mission. employees are more likely to be motivated and engaged in CSR efforts when they see a clear connection between their daily work and the organization's social impact.

Establish a Cross-Functional CSR Team: Form a dedicated cross-functional CSR team comprising representatives from different departments. This team can collaborate on planning and executing CSR projects, ensuring diverse perspectives and expertise are included.

Provide Opportunities for Employee Input: Include employees in the CSR initiative decision-making process. Conduct surveys, focus groups, or town hall meetings to collect employee feedback and suggestions. This gives them the ability to shape CSR efforts and a sense of ownership.

Volunteer and Skill-Based Programs: Provide employees with opportunities to donate their time and expertise to CSR initiatives. In addition, devise skill-based volunteer programmes in which employees can contribute their professional skills to social causes. This can result in a more fulfilling and meaningful engagement experience.

Recognize and Celebrate Contributions: Recognise and honour the contributions of employees who participate actively in CSR initiatives. Utilise internal communications, newsletters, or employee profiles to

   The Comprehensive Guide On CSR For Consultant

acknowledge their efforts. Publicly recognising their contribution can encourage others to participate as well.

Communicate Impact : Share the impact of CSR initiatives regularly with employees. Provide updates on the progress and outcomes of CSR initiatives, highlighting how their efforts have impacted the community or environment. Transparent communication assists employees in comprehending the significance of their participation.

Organize CSR Events and Activities: Organize events and activities related to CSR initiatives that encourage employee participation. These could include charity drives, environmental clean-ups, or fundraising campaigns. Such events foster team spirit and a shared commitment to social responsibility.

Employee Recognition Program: Implement a programme to recognise employees for their contributions to corporate social responsibility. Consider creating awards or certificates for exceptional CSR efforts in order to motivate and inspire others to participate actively in social initiatives.

Training and Education: Provide training and educational sessions on CSR, sustainability, and social issues. This helps employees develop a deeper understanding of the impact of their actions and enables them to become better advocates for positive change.

Support Employee-Led Initiatives: Encourage employees to initiate their own CSR projects or to participate in existing external initiatives. Providing support and resources for these employee-led initiatives demonstrates the organization's dedication to employee-driven social responsibility.

 The Comprehensive Guide On CSR For Consultant

By implementing these strategies, organisations can cultivate a culture of employee engagement in CSR initiatives, resulting in a workplace that is more socially responsible and mission-driven. Engaged employees become advocates for the company's CSR initiatives, ultimately contributing to a positive social and environmental impact.

Example: Please check the annexure for detail example

10.9: Corporate Social Responsibility Reporting

It is necessary to include both qualitative and quantitative aspects of the program's inputs, outputs, and outcomes when writing a comprehensive report on Corporate Social Responsibility (CSR) social development initiatives. This will offer a comprehensive view of CSR initiatives and their effects. How to write about these elements in your CSR report is outlined below.

<table>
<tr><td>

Introduction:
Provide a Brief introduction to the CSR initiatives, highlighting the company's commitment to social development and the report's objective.
Qualitative Inputs:
- Describe the qualitative resources and efforts invested in the CSR initiatives.
- Highlight the commitment of senior management to CSR decision-making and the expertise of the CSR team and volunteers.
- Explain any strategic partnerships or collaborations with NGOs, government agencies, and local communities.
- Discuss how stakeholder engagement and feedback mechanisms were incorporated to shape

</td></tr>
</table>

   The Comprehensive Guide On CSR For Consultant

the initiatives.

Quantitative Inputs:

- Present the measurable resources or contributions invested in the CSR initiatives.
- Provide the financial investments, total monetary contributions, and the budget allocated for the initiatives.
- State the number of volunteer hours dedicated to CSR projects and any other quantitative resources provided.

Qualitative Outputs:

- Describe the immediate and tangible results of the CSR initiatives.
- Explain how the initiatives have raised awareness and knowledge among the targeted beneficiaries.
- Highlight the level of community engagement and participation achieved through the CSR events.
- Include any positive feedback and testimonials from beneficiaries and stakeholders.

Quantitative Outputs:

- Present the measurable and observable results of the CSR initiatives.
- State the number of beneficiaries reached by the CSR programs.
- Provide percentage data or numerical values indicating the adoption of sustainable practices resulting from the initiatives.
- Include the quantity of goods or services distributed to the target population.

Qualitative Outcomes:

- Describe the broader, long-term changes or impacts resulting from the CSR initiatives.

The Comprehensive Guide On CSR For Consultant

- Explain how the initiatives have improved the quality of life for beneficiaries and communities.
- Highlight any evidence of increased social cohesion and community empowerment.
- Discuss the impact on brand reputation and stakeholder trust.

Quantitative Outcomes:

- Present the measurable and objective changes resulting from the CSR initiatives.
- Provide data on reductions in carbon emissions or waste generation achieved through the initiatives.
- Include statistics indicating increases in literacy rates or educational attainment levels among the target population.
- Mention any quantitative evidence of reduced poverty levels among the beneficiaries.

Case Studies and Success Stories:

- Include specific case studies or success stories that illustrate the qualitative and quantitative impacts of individual CSR initiatives.
- Use narratives, data, and testimonials to showcase the success of the initiatives.

Challenges and Lessons Learned:

- Acknowledge any challenges faced during the implementation of the CSR initiatives.
- Explain the lessons learned from overcoming these challenges.

Conclusion:

Summarize the overall impact and effectiveness of the CSR social development initiatives.

Express gratitude to all stakeholders, partners, and volunteers for their contributions to the success of the initiatives.

   The Comprehensive Guide On CSR For Consultant

By incorporating both qualitative and quantitative aspects in your CSR report, you can provide a comprehensive and data-driven assessment of the social development initiatives, showcasing the positive changes brought about by the company's efforts in creating a sustainable and positive impact on society.

FORMAT:

Corporate Social Responsibility (CSR) Report: [Company Name]
Introduction:
Under the Section 135 of the Companies Act, 2013, and the Companies (Corporate Social Responsibility Policy) Rules, 2014 require [Company Name] to fulfil its social, environmental, and ethical responsibilities. This report details our CSR initiatives for the fiscal year [Year], which reflect our efforts to positively affect society and the environment.

CSR Policy and Governance:
Our CSR Committee, comprised of three directors [Name & Designation], one of whom is independent, is responsible for formulating, implementing, and directing our CSR initiatives. The committee has developed a comprehensive CSR policy consistent with the relevant provisions of the Companies Act and our company's core values. [CSR Policy Link]

CSR Expenditure:
During the financial year [Year], our company's net profit was [Amount]. Accordingly, we allocated [2% of the average net profit of the preceding three financial

years] towards CSR initiatives, resulting in an expenditure of [Amount] during the reporting period.

Key CSR Areas and Initiatives:
Eg.1. Education and Skill Development:
We partnered with local educational institutions to enhance access to quality education for underprivileged children.
We conducted vocational training programs to empower youth with employable skills.
Eg 2. Healthcare and Sanitation:
We supported healthcare centers in rural areas to provide medical facilities and awareness programs.
We contributed to sanitation projects, promoting hygiene and access to clean water.
Eg. 3. Environmental Sustainability:
Our company implemented energy-saving measures, reducing our carbon footprint.
We engaged in afforestation drives and biodiversity conservation efforts.
Eg. 4. Women Empowerment:
We launched initiatives to promote gender equality and provide opportunities for women's economic empowerment.
We supported self-help groups and women entrepreneurs through capacity-building programs.
Eg. 5. Rural Development:
Our company invested in community infrastructure development projects, such as roads and schools.
We promoted agricultural practices and livelihood enhancement programs for farmers.
CSR Impact Assessment:
We believe in measuring the results of our CSR efforts. During the fiscal year [Year], we conducted independent evaluations of the effectiveness and viability of our

 The Comprehensive Guide On CSR For Consultant

initiatives. Our impact assessment revealed positive results in a variety of areas, including an increase in school enrollment, enhanced access to healthcare, and improved livelihoods for rural communities.[CSR Impact Assessment Report Link]

Stakeholder Engagement:
Our organisation acknowledges the significance of involving stakeholders in the CSR process. To align our initiatives with the requirements and priorities of the beneficiaries, we solicited feedback and suggestions from local communities, employees, NGOs, and government authorities.

Future Plans:
As part of our commitment to CSR over the long term, we intend to expand present initiatives and probe new intervention areas. We intend to increase our emphasis on education, healthcare, and environmental sustainability, while continuously seeking opportunities to have a significant social impact.

Conclusion:
[Company Name] is proud of its CSR achievements in the financial year [Year]. We remain steadfast in our dedication to responsible business practices, sustainable development, and making a lasting difference in the communities we serve. We express our gratitude to all stakeholders, partners, and employees for their unwavering support in our CSR journey.

[Signature of Authorized Representative]
[Name and Designation]

The Comprehensive Guide On CSR For Consultant

10.10: Corporate Social Responsibility (CSR) Audit

Corporate Social Responsibility (CSR) audits are based on the principle of assessing and evaluating an organization's CSR practises, performance, and impact. This audit ensures the organization's CSR initiatives align with its specified values, objectives, and commitments. The CSR audit adheres to these guidelines:

Conducting a comprehensive assessment of the organization's CSR-related activities, policies, and procedures. This includes a review of the organization's environmental practises, social initiatives, stakeholder engagement, supply chain management, employee welfare, and compliance with applicable laws and regulations. In order to gain a comprehensive understanding of the organization's CSR performance, the evaluation should include data collection, interviews, and document review.

Transparency and Accountability in Reporting: The CSR audit should emphasise transparency and accountability in reporting. This includes establishing clear criteria and standards for evaluating the CSR performance of the organisation and ensuring that the audit process is conducted objectively and independently. The audit findings should be reported in a manner that accurately reflects the organization's CSR strengths, weaknesses, and advancement opportunities. The organisation should be accountable for addressing identified deficiencies, implementing corrective actions, and informing stakeholders of progress.

 The Comprehensive Guide On CSR For Consultant

By adhering to these principles, organisations are able to conduct a comprehensive CSR audit that evaluates the efficacy of their CSR practises, identifies areas for improvement, and promotes accountability. The CSR audit assists organisations in aligning their actions with their declared CSR commitments, promotes continuous improvement, and increases stakeholders' confidence and credibility.

The process of conducting a CSR audit may vary depending on the organization's size, industry, and scope of CSR activities. Here's an hypothetical example of a Corporate Social Responsibility (CSR) audit for a fictional company called "EcoTech Solutions ltd," which is a technology company committed to environmental sustainability and social impact. However, some common steps in a CSR audit typically include:

CSR Audit :"EcoTech Solutions ltd,"

Defining Scope and Objectives: Clearly define the scope of the CSR audit, including the specific areas and initiatives to be assessed. Establish the audit objectives and what the organization aims to achieve through the audit.

Example: The scope of the CSR audit for EcoTech Solutions ltd includes assessing the company's environmental initiatives, social programs, and ethical practices. The objectives are to evaluate the effectiveness of current CSR efforts, identify opportunities for improvement, and ensure alignment with the company's sustainability goals.

 The Comprehensive Guide On CSR For Consultant

Reviewing Policies and Documentation: Examine the organization's CSR policies, codes of conduct, sustainability reports, and other relevant documentation to understand its commitments and initiatives.

Example: The audit team examines EcoTech Solutions ltd CSR policy, code of conduct, sustainability reports, and any other relevant documents related to CSR commitments.

Stakeholder Engagement: Involve key stakeholders, both internal (employees, management, etc.) and external (customers, suppliers, communities, NGOs, etc.), to gather their perspectives on the organization's CSR efforts.

Example: The audit team conducts interviews with employees, customers, suppliers, and local communities to gather their perspectives on the company's CSR initiatives and their impact.

Assessing CSR Implementation: Evaluate how the organization translates its CSR commitments into action. This may involve reviewing specific projects, programs, and initiatives undertaken by the organization.

Example: The team evaluates specific CSR projects, such as:

Eco-friendly product development and sustainable design practices.

Reduction of greenhouse gas emissions and energy consumption in company operations.

Community engagement programs focused on digital literacy and skills development for underprivileged youth.

Measuring Impact: Analyze the organization's CSR performance metrics and data to assess the impact of its initiatives on social, environmental, and economic aspects.

Example: The team analyzes data on the following CSR metrics:

- The percentage reduction in the company's carbon footprint over the past three years.
- The number of students enrolled in the digital literacy programs and their employment outcomes.
- The percentage of suppliers complying with EcoTech's Ltd ethical sourcing guidelines.

Identifying Gaps and Opportunities: Identify areas where the organization is excelling in CSR and areas that require improvement or further attention.

Example: The audit identifies areas where EcoTech excels, such as its innovative eco-friendly products. However, it also highlights opportunities to expand community outreach efforts and engage more suppliers in sustainable practices.

Risk Assessment: Evaluate potential risks associated with the organization's CSR activities, including reputational risks, compliance risks, and environmental risks.

Example: The audit team assesses potential risks, such as reputational risks associated with any non-compliance with environmental regulations.

Comparative Analysis: Benchmark the organization's CSR performance against industry peers and best practices.

Example: The team benchmarks EcoTech's CSR performance against industry peers and best practices to gain insights into how it compares with other companies

 The Comprehensive Guide On CSR For Consultant

in the sector.

Compliance Check: Ensure that the organization is compliant with relevant laws, regulations, and reporting requirements related to CSR.
Example: The audit verifies EcoTech's compliance with relevant environmental regulations and reporting requirements.

Recommendations and Action Plan: Based on the findings of the audit, provide recommendations for enhancing the organization's CSR strategy and performance. Develop an action plan to address identified gaps and capitalize on opportunities.
Example: Based on the audit findings, the team provides the following recommendations:

- Increase investment in community programs to reach a larger number of beneficiaries.
- Develop a formal supplier engagement program to promote sustainable sourcing practices.
- Enhance internal monitoring and reporting mechanisms to track CSR performance more effectively.

Reporting and Communication: Prepare a comprehensive CSR audit report, summarizing the audit process, findings, and recommendations. Communicate the results to relevant stakeholders, including management, employees, and external partners.
Example: The CSR audit report is prepared, including a summary of the audit process, findings, and detailed recommendations. It is shared with EcoTech's senior management and CSR team.

Continuous Improvement: A CSR audit is not a one-time exercise. Encourage the organization to use the audit as a foundation for continuous improvement in its CSR efforts.

Example: EcoTech uses the audit results to update its CSR strategy, set new goals, and integrate sustainability principles further into its business operations. The company commits to regular audits to track progress and continuously improve its CSR performance.

By conducting a comprehensive CSR audit like this, EcoTech Solutions gains valuable insights into its CSR initiatives' effectiveness and can make informed decisions to strengthen its commitment to social responsibility and sustainability.

A well-conducted CSR audit helps organizations gain insights into their social and environmental impact, enhance transparency, and build trust with stakeholders. It can also serve as a basis for setting meaningful CSR goals and integrating sustainability into the organization's overall business strategy.

CHAPTER 11: THE NEED: SOCIAL MEDIA MANAGEMENT

Social media management is crucial for both Corporate Social Responsibility (CSR) teams and Non-Governmental Organizations (NGOs) for several reasons:

Visibility and Awareness	The vast user base of social media platforms makes them ideal for raising awareness about CSR initiatives and NGO causes. Effective management of social media can increase the visibility of initiatives and attract a larger audience.
Engagement and Interaction	CSR teams and NGOs can engage directly with their audience, including beneficiaries, donors, and partners, as well as the general public, through social media. It allows for two-way communication, which fosters a sense of community and participation.
Brand Building	Consistent and strategic social media management can contribute to the development of a positive brand image for the organisation and its CSR initiatives. It highlights the organization's dedication to social

 The Comprehensive Guide On CSR For Consultant

	causes and enhances its standing.
Storytelling	Social media provides a platform for sharing impactful stories, testimonies, and success stories related to CSR projects. Storytelling humanizes the projects and creates emotional connections with the audience.
Fundraising and Donor Engagement	For NGOs, social media can be a powerful tool for fundraising. By showcasing their projects and impact, NGOs can attract potential donors and engage existing ones.
Real-time Updates	Social media enables organizations to provide real-time updates on project activities, milestones, and events. This level of transparency enhances trust and credibility among stakeholders.
Advocacy and Policy Influence	Both CSR teams and NGOs can leverage social media to advocate for social and environmental issues, influencing public opinion and policymakers.
Crowdsourcing and Volunteer Engagement	Social media platforms can be used to mobilize volunteers, invite community participation, and seek support for various activities and campaigns.
Measuring Impact	Social media analytics provide valuable insights into audience engagement, reach, and sentiment. These metrics help organizations understand the effectiveness of their communication strategies and make

	data-driven decisions.
Cost-Effectiveness	Social media is a cost-effective way to reach a vast audience compared to traditional advertising or outreach methods.
Global Reach	Social media transcends geographical boundaries, enabling CSR teams and NGOs to engage with a global audience and potential partners.
Crisis Management	In times of crises or emergencies, social media can be used to disseminate vital information, address concerns, and provide support.
Collaboration and Partnerships	Social media can facilitate collaborations with like-minded organizations, influencers, and change-makers, fostering a supportive network for social impact.

To maximise the benefits of social media management, CSR teams and NGOs should establish a well-defined social media strategy, adapt their messaging to the platform and audience, and communicate in an authentic and consistent manner. Embracing the power of social media can substantially amplify the impact of CSR initiatives and increase support for causes championed by NGOs.

 The Comprehensive Guide On CSR For Consultant

CHAPTER 12: DIFFICULTIES FACING CORPORATE SOCIAL RESPONSIBILITY

Corporate social responsibility (CSR) initiatives face several challenges that organizations must navigate to successfully implement and sustain their social and environmental commitments. Some of the key challenges faced by CSR are:

Limited consumer demand: In certain markets or industries, consumer awareness and demand for socially responsible products and services may be limited. Companies attempting to invest in and promote CSR initiatives may face obstacles due to the lack of significant consumer demand. To surmount this obstacle, it is essential to educate and motivate consumers about the value and advantages of making responsible decisions.

Difficulty in engaging small and medium-sized enterprises (SMEs): Engaging SMEs in CSR initiatives can be challenging due to resource constraints, limited awareness, and lack of expertise. SMEs may face difficulty in implementing and reporting on CSR practices, as they often have fewer resources and dedicated CSR teams. Supporting and facilitating CSR adoption among SMEs is important for achieving comprehensive and inclusive social responsibility.

Changing stakeholder expectations: Stakeholder expectations around CSR can evolve rapidly, and organizations must stay attuned to these changes. The diverse range of stakeholders, including customers, employees, investors, and communities, may have varying and evolving expectations for CSR. Organizations need to continuously engage with stakeholders, understand their evolving needs, and adapt their CSR initiatives accordingly.

Alignment with business objectives: Achieving a balance between social and environmental goals and core business objectives can be challenging. Organizations must ensure that CSR initiatives align with their overall business strategies and financial goals. Demonstrating the business case and tangible benefits of CSR, such as improved brand reputation, employee engagement, and risk management, is crucial for organizational buy-in.

Lack of collaboration and knowledge sharing: Collaboration among organizations, NGOs, and other stakeholders is essential for addressing complex social and environmental challenges effectively. However, barriers to collaboration, such as competition, lack of trust, and limited knowledge sharing, can hinder progress. Building collaborative networks, sharing best practices, and fostering a culture of cooperation are necessary to overcome this challenge.

Long-term commitment and sustainability: Sustaining CSR initiatives over the long term requires ongoing commitment and investment from organizations. It can be challenging to maintain momentum, secure financial resources, and prioritize CSR amid changing market conditions and competing business priorities. Organizations must embed CSR into their core values and corporate culture to ensure long-term sustainability.

 The Comprehensive Guide On CSR For Consultant

Resistance to change: Implementing CSR initiatives often requires changes in organizational culture, processes, and practices. Resistance to change can arise from employees, management, or other stakeholders who may be resistant to new ways of doing things or perceive CSR as an additional burden or cost. Overcoming resistance and fostering a culture of CSR within the organization can be a challenge.

Lack of standardization and measurement: There is a lack of standardized frameworks and metrics for measuring and reporting CSR performance. This makes it challenging to compare and benchmark CSR efforts across industries and organizations. Lack of standardized measurement can also make it difficult for organizations to set goals, track progress, and communicate their CSR achievements effectively.

Limited resources for implementation: Allocating resources, including financial, human, and technological, to CSR initiatives can be a challenge, particularly for smaller or resource-constrained organizations. Limited resources can impede the implementation of comprehensive CSR programs or hinder organizations from addressing complex social and environmental issues adequately.

Complexity of global operations: Multinational corporations operating in multiple countries face the challenge of navigating diverse legal, cultural, and social contexts. Adapting CSR initiatives to different regions and ensuring compliance with local regulations and cultural sensitivities can be complex. Organizations must develop strategies to manage the complexities of their global operations while maintaining consistent CSR standards.

 The Comprehensive Guide On CSR For Consultant

Greenwashing and skepticism: The phenomenon of "greenwashing," where organizations make misleading or exaggerated claims about their environmental or social responsibility, is a challenge for CSR. Skepticism from stakeholders and the public can arise due to past instances of greenwashing or a perception that CSR initiatives are merely PR tactics. Building trust and credibility through genuine and transparent CSR efforts is crucial.

Balancing short-term financial goals with long-term sustainability: Balancing short-term financial goals with long-term sustainability objectives can be challenging. Organizations may face pressures to prioritize immediate financial gains over long-term investments in CSR. Striking the right balance between financial performance and sustainable practices is crucial for the success and credibility of CSR initiatives.

Cultural and geographical differences: Operating in diverse cultural and geographical contexts presents challenges for CSR initiatives. What may be considered socially responsible in one region may not align with local norms or expectations in another. Organizations need to navigate these differences, adapt their initiatives, and engage with local stakeholders to ensure cultural sensitivity and relevance.

Engaging the supply chain: Extending CSR practices throughout the supply chain can be a significant challenge, especially in industries with complex and global supply networks. Ensuring that suppliers and subcontractors adhere to CSR standards and ethical practices requires robust monitoring and collaboration. Engaging and

 The Comprehensive Guide On CSR For Consultant

incentivizing the supply chain to align with CSR goals can be a complex task.

Complexity of supply chains: Many organizations operate complex global supply chains, which can present challenges in ensuring social and environmental responsibility throughout the entire chain. Managing and monitoring suppliers, subcontractors, and partners for compliance with CSR standards can be difficult, especially in industries with extensive networks and high-risk sourcing regions.

Time and patience: Implementing effective CSR initiatives often requires long-term commitments and patience. Some social and environmental issues cannot be resolved quickly, and achieving meaningful impact may take time. Organizations need to remain dedicated and resilient in their CSR efforts, continuously learning and adapting their approaches to address persistent challenges.

Balancing stakeholder interests: Balancing the diverse interests of stakeholders can be a complex task. Different stakeholders may have conflicting priorities, making it challenging to satisfy everyone's expectations. Organizations must navigate these competing interests while maintaining their commitment to ethical practices and sustainability.

Limited awareness and education: Lack of awareness and understanding of CSR among employees, customers, and the general public can be a challenge. Educating stakeholders about the importance and benefits of CSR, as well as the organization's specific initiatives, is crucial for building support and engagement. Effective communication

  The Comprehensive Guide On CSR For Consultant

and awareness campaigns are essential to overcome this challenge.

Political and economic volatility: CSR initiatives can be influenced by political and economic factors beyond the control of organizations. Changes in government regulations, economic downturns, or political instability can impact the sustainability and continuity of CSR programs. Organizations need to be prepared to adapt their strategies and activities in response to external changes.

Limited collaboration and partnerships: Addressing complex social and environmental issues often requires collaboration among various stakeholders, including governments, NGOs, communities, and other businesses. However, building effective partnerships and collaborations can be challenging due to competition, conflicting interests, and resource limitations. Organizations need to foster a collaborative mindset and establish mutually beneficial relationships to overcome this challenge.

Scaling up and replication: Scaling up successful CSR initiatives or replicating them across different locations or business units can be a challenge. Adapting programs to different contexts and ensuring consistency in implementation while allowing for local customization requires careful planning and coordination. Organizations must develop strategies to scale and replicate their CSR initiatives effectively.

Evolving societal expectations: Societal expectations around CSR continue to evolve, and organizations must keep pace with these changes. As issues such as diversity and inclusion, climate change, and human rights gain

   The Comprehensive Guide On CSR For Consultant

prominence, organizations need to continuously reassess their CSR strategies and initiatives to address emerging concerns effectively.

Lack of clear strategy and integration: Developing a clear CSR strategy that aligns with the organization's core values, mission, and business objectives can be challenging. Many companies struggle with integrating CSR initiatives into their overall business strategy, resulting in disjointed efforts and limited impact. Lack of strategic alignment can lead to CSR programs being perceived as superficial or disconnected from the organization's core operations.

Resource constraints: Implementing robust CSR initiatives often requires significant financial, human, and technological resources. Small and medium-sized enterprises (SMEs) may face particular challenges in allocating resources to CSR due to limited budgets and capacity. Resource constraints can hinder the development and implementation of comprehensive CSR programs, making it difficult for organizations to address complex social and environmental issues effectively.

Measuring impact and ROI: Quantifying the impact of CSR initiatives and demonstrating a return on investment (ROI) can be complex. It can be challenging to establish clear cause-and-effect relationships between CSR activities and tangible outcomes, making it difficult to measure the true impact of these initiatives. Additionally, measuring intangible social and environmental benefits can be subjective and require innovative methodologies.

   The Comprehensive Guide On CSR For Consultant

Stakeholder engagement and expectations: Engaging diverse stakeholders and managing their expectations is a significant challenge for CSR initiatives. Stakeholders may have varying priorities, interests, and expectations, which can be difficult to reconcile. Balancing stakeholder interests while ensuring the initiative's integrity and alignment with the organization's values requires effective communication, collaboration, and relationship management.

Addressing complex social and environmental issues: CSR initiatives often aim to tackle complex social and environmental challenges, such as poverty, climate change, or human rights violations. These issues typically require multi-faceted approaches and collaboration across sectors to achieve meaningful impact. Navigating complex problems and finding sustainable solutions can be challenging, requiring organizations to leverage partnerships, expertise, and innovative approaches.

Reputation and credibility risks: Companies engaging in CSR initiatives face reputation and credibility risks if their actions are perceived as insincere or greenwashing. Stakeholders increasingly expect organizations to demonstrate transparency, authenticity, and genuine commitment to social and environmental causes. Lack of transparency, inconsistent messaging, or failure to deliver on CSR commitments can lead to reputational damage and loss of stakeholder trust.

Regulatory and legal complexities: CSR initiatives operate within a complex legal and regulatory landscape. Companies must navigate diverse and evolving regulatory frameworks that govern social and environmental practices. Complying with local, national, and international

 The Comprehensive Guide On CSR For Consultant

regulations while aligning with best practices and voluntary standards can be challenging, especially for multinational corporations operating in different jurisdictions.

Cultural and contextual differences: Companies operating globally encounter cultural and contextual challenges when implementing CSR initiatives across diverse markets and communities. Understanding local customs, values, and societal norms is essential for designing and implementing effective CSR programs that are respectful and relevant. Failure to consider cultural and contextual differences can lead to misunderstandings, resistance, and ineffective outcomes.

Overcoming these challenges requires strong leadership, commitment, and a systematic approach to CSR. Organizations need to develop clear strategies, allocate appropriate resources, engage stakeholders, measure and communicate impact, and navigate regulatory complexities. By addressing these challenges proactively, companies can enhance the effectiveness and sustainability of their CSR initiatives, maximizing their positive social and environmental contributions.

Annexure 1: Structure for Baseline Study Report Writing in Corporate Social Responsibility (CSR)

A baseline study is a crucial component of corporate social responsibility (CSR) initiatives as it provides a starting point for measuring the impact of the CSR project. The baseline study report serves as a foundation to assess changes and improvements resulting from the CSR intervention over time. Here's a detailed Structure for writing a comprehensive baseline study report in reference to CSR:

Executive Summary:

- A concise overview of the CSR project and its objectives.
- Summary of key findings from the baseline study.
- Major challenges and opportunities identified.
- Summary of recommended actions based on the findings.

Introduction:

- Background information about the CSR project and its purpose.
- Objectives of the baseline study and its significance in evaluating the impact of the CSR initiative.

The Comprehensive Guide On CSR For Consultant

- Description of the target beneficiaries and the geographic area covered.

Literature Review:

- Review relevant literature and existing research on CSR in the organization's industry or sector.
- Summarize key theories, frameworks, and best practices related to CSR implementation and impact assessment.
- Identify gaps or areas requiring further investigation in the existing literature.

Methodology:

- Explanation of the research design and approach used for the baseline study.
- Details about the data collection methods, including surveys, interviews, focus groups, and secondary data sources.
- Information about the sample size and selection criteria.
- Description of any limitations or constraints faced during data collection.

Socioeconomic and Demographic Profile:

- Detailed profile of the target community or beneficiaries.
- Information on demographics, such as age, gender, education levels, income, and occupation.
- Overview of the community's socioeconomic status and living conditions.

Needs Assessment:

 The Comprehensive Guide On CSR For Consultant

- Identification of the key needs, challenges, and problems faced by the target beneficiaries.
- Presentation of evidence and data supporting the identified needs.
- Analysis of the urgency and severity of the identified needs.

Baseline Indicators:

- List of baseline indicators relevant to the CSR project's objectives.
- Clear definition of each indicator and how it will be measured over time.
- Presentation of baseline data for each indicator.

Stakeholder Analysis:

- Identification of the stakeholders involved in the CSR project.
- Assessment of their interests, influence, and level of involvement.
- Evaluation of stakeholder support and collaboration during the baseline study.

Data Analysis and Findings

- Present and analyze the baseline data collected, including quantitative and qualitative information.
- Discuss the findings in relation to the organization's CSR goals, objectives, and performance targets.
- Identify gaps, strengths, and areas for improvement in the organization's CSR practices. SWOT Analysis:

Conduct a SWOT (Strengths, Weaknesses, Opportunities, Threats) analysis specific to the organization's CSR efforts.

- Identify the internal and external factors that impact the organization's CSR performance.
- Assess the strengths to leverage, weaknesses to address, opportunities to pursue, and threats to mitigate.

Key Challenges and Opportunities:

- Identification and discussion of the major challenges and obstacles encountered during the baseline study.
- Exploration of opportunities and potential areas of growth or collaboration.

Implementation Plan:

- Develop a detailed implementation plan for the recommended CSR initiatives.
- Define specific actions, responsible parties, timelines, and resource requirements.
- Outline monitoring and evaluation mechanisms to track progress and measure impact.

Recommendations:

- Actionable recommendations based on the findings and analysis.
- Strategies to address the identified challenges and leverage opportunities.
- Proposed steps to enhance the effectiveness and impact of the CSR initiative.

 The Comprehensive Guide On CSR For Consultant

Conclusion:

- Summarization of the main points discussed in the baseline study report.
- Reiteration of the importance of the baseline study in guiding the CSR project's implementation and evaluation.
- Emphasis on the value of using the baseline data as a reference for future assessments.

Annexes:

- Inclusion of additional information, questionnaires, and data collection tools used during the baseline study.
- Appendices for supplementary data, charts, and graphs.

References:

- Listing of all the sources and references cited throughout the baseline study report.

By adhering to this Structure, the baseline study report will be well-organized, insightful, and serve as a valuable reference for the CSR project's progress and impact assessment over time.

Basic questionnaire of Baseline Survey

A baseline survey for a social development project typically covers various thematic areas relevant to the project's objectives. Below is a detailed questionnaire that encompasses different thematic areas:

Section 1: Respondent Information

- What is your name? (Optional)

- Gender:
- Age:
- Contact information: (Optional)
- Phone number:
- Email:

Section 2: Socioeconomic Background

- Education level:
- Occupation:
- Household size:
- Monthly household income:
- Type of housing:
- Own house
- Rented house
- Shared accommodation
- Other (please specify):

Section 3: Community and Social Environment

- How long have you been living in this community?
- What are the main strengths or positive aspects of this community?
- What are the main challenges or issues faced by this community?
- How would you describe the overall sense of community and social cohesion in this area?

Section 4: Needs and Priorities

1. **In your opinion, what are the most pressing social needs in this community? (Select up to**

 The Comprehensive Guide On CSR For Consultant

three)
- Access to quality education
- Healthcare services
- Employment opportunities
- Affordable housing
- Poverty alleviation
- Environmental sustainability
- Gender equality
- Community safety and security
- Social inclusion and diversity
- Other (please specify):

2. How would you rank the selected social needs in terms of priority? (1 = highest priority, 3 = lowest priority)

Section 5: Existing Programs and Services

1. Are you aware of any existing programs or services addressing the identified social needs in this community? (Select all that apply)

- Education programs
- Health clinics or hospitals
- Job training or employment assistance
- Affordable housing initiatives
- Poverty alleviation programs
- Environmental conservation projects
- Gender equality initiatives
- Community safety programs
- Social inclusion and diversity programs
- Other (please specify):

2. How effective do you think these existing

programs or services are in addressing the identified social needs? (Scale of 1 to 5, with 1 being not effective and 5 being highly effective)

Section 6: Project-Specific Questions

- Are you aware of the social development project/initiative being implemented in this community?
- What do you perceive as the main objectives or goals of this project/initiative?
- How do you think this project/initiative will contribute to addressing the social needs and challenges in this community?

Section 7: Communication and Engagement

1. **How would you prefer to receive information and updates about the project/initiative?** (Select all that apply)
- Community meetings
- Newsletters or pamphlets
- Social media platforms
- Text messages or emails
- **Local radio or TV announcements**
- **Other (please specify):**

2. **How interested would you be in actively participating in the project/initiative? (Scale of 1 to 5, with 1 being not interested and 5 being highly interested)**

3. **Would you be interested in actively participating in the project/initiative? If yes, in what capacity? (Select all that apply)**

- Volunteering
- Providing input and feedback
- Joining a community committee or group
- Attending workshops or training sessions
- Other (please specify):

Section 8: Additional Comments

- Do you have any additional comments, suggestions, or concerns regarding the social development project or the identified social needs in this community?

Thank the respondent for their participation and assure them that their responses will be kept confidential and used only for research and project planning purposes.

Ensure that the questionnaire is administered in a culturally sensitive and respectful manner. Pilot test the survey with a small group of participants to identify any potential issues before implementing it on a larger scale. Moreover, consider translating the questionnaire into local languages if necessary to ensure inclusivity and accurate responses.

   The Comprehensive Guide On CSR For Consultant

Annexure 2 : Logical Structure Analysis

Logical Structure Analysis, also known as Logframe Analysis or Logframe Approach, is a structured and methodical planning and management instrument used to design, monitor, and evaluate development projects, programmes, and interventions. It provides a succinct and logical representation of the project's goals, objectives, activities, outputs, outcomes, and indicators, as well as verification methods and assumptions. In the context of development and aid programmes, as well as Corporate Social Responsibility (CSR) initiatives, Logframe Analysis is widely utilised.

The Logframe consists of four main components:

1. **Goal (Overall Objective):**
   - The highest-level objective of the project, reflecting its ultimate purpose and intended impact.
   - It should be broad, ambitious, and aligned with the project's mission.
2. **Purpose (Specific Objectives):**

- Specific and measurable objectives that directly contribute to achieving the project's goal.
- They are the outcomes that the project aims to achieve.

3. **Outputs:**
   - Concrete and measurable deliverables or results produced by the project activities.
   - Outputs are the immediate results of the project's efforts.

4. **Activities:**
   - Specific actions and interventions that the project will undertake to produce the desired outputs.
   - Activities are the actions that the project team will implement.

Each component is linked through cause-and-effect relationships, creating a logical chain of events that leads from the activities to the desired impact. The Logframe helps stakeholders understand the project's theory of change and provides a clear roadmap for project planning, implementation, and evaluation.

Logframe Matrix:

The Logframe is commonly presented in a matrix format, also known as the Logical Structure Matrix. It includes the following columns:

1. **Vertical Logic (Hierarchy):**
   - Goal
   - Purpose
   - Outputs
   - Activities

 The Comprehensive Guide On CSR For Consultant

2. **Horizontal Logic (Indicators, Means of Verification, and Assumptions):**

- Indicators: Quantifiable and measurable parameters that will be used to track progress and success.
- Means of Verification (MoV): The sources and methods to collect data for each indicator.
- Assumptions: External factors and conditions that could influence the project's success but are beyond its control.

Level	Component	Indicator	Means of Verification	Assumptions
Goal	Improve Educational Assess in the Rural Community	Increase Assess to Quality Education	Government report and statistics	Government Support for Education
Purpose	Establish Community Learning Centers	Number of Community Leaner Center	Projects Reports and Records	Community willingness to participate
Output	Train Local Teacher	Number of Local Teacher Trained	Training Attendance	Availability of Qualified trainers
Output	Provide Learning Material	Increased availabili	Monitoring Reports	Availability of Learning

		ty of learning material	and Beneficiary feedback	material
Activity	Conduct Teacher Training workshops	Number of Teachers Workshops	Training Attendance	Availability of Qualified trainers
Activity	Distribute Learning material	Number of Learning Material	Distribution Records	Availability of learning Material

In this example, the Logframe Matrix outlines the various components of a project aimed at improving education access in rural communities. The Logframe provides a clear structure for project planning and serves as a basis for monitoring and evaluation activities.

By using Logical Structure Analysis, stakeholders can ensure that their projects are well-designed, feasible, and aligned with the desired outcomes and objectives. It also facilitates transparency, accountability, and effective communication among project stakeholders.

The Comprehensive Guide On CSR For Consultant

Annexure 3: Corporate Social Responsibility Proposal

Developing a well-structured NGO proposal is necessary for fundraising success. The following are the steps to creating an effective proposal: Recognize the Funder's Requirements, Investigate potential sponsors and become familiar with their priorities, areas of interest, and financing rules.

A well-structured and compelling NGO proposal is crucial for successful fundraising efforts. Here's a Structure for an NGO proposal for fund raising:

1. **Cover Page:**
   - Include the NGO's name, logo, and contact details.
   - Mention the title of the proposal and the date.
2. **Executive Summary:**
   - Provide a concise overview of the NGO's mission, objectives, and the project for which funding is sought.
   - Highlight the key aspects of the proposal, including the project's goals and the amount of funding required.

3. **Introduction:**
   - Introduce the NGO and its history, including its vision and mission.
   - Provide background information about the NGO's past projects and achievements.
   - Explain the purpose of the proposal and the specific project it aims to fund.

4. **Needs Assessment:**
   - Clearly outline the problem or issue the NGO seeks to address with the proposed project.
   - Present data, statistics, and evidence that demonstrate the significance of the problem.
   - Highlight the beneficiaries and communities that will be impacted by the project.

5. **Project Description:**
   - Detail the objectives, activities, and outcomes of the proposed project.
   - Explain how the project will address the identified needs and contribute to the NGO's mission.
   - Provide a timeline for project implementation and the expected duration.

6. **Budget and Financial Plan:**
   - Present a detailed budget for the entire project, including itemized expenses for each activity.
   - Specify the amount of funding requested and how it will be utilized.
   - Include information on any existing funding or resources available for the project.

7. **Monitoring and Evaluation Plan:**
   - Describe the methods and indicators that will be used to measure the project's progress and success.

 The Comprehensive Guide On CSR For Consultant

- Explain how the NGO will track the use of funds and ensure transparency and accountability.

8. **Sustainability Plan:**
   - Demonstrate how the project will be sustained after the funding period ends.
   - Discuss strategies for generating additional resources or partnerships.

9. **Organizational Capacity:**
   - Showcase the NGO's capabilities, experience, and expertise in implementing similar projects.
   - Provide information about the NGO's staff, board members, and key partners.

10. **Impact and Benefits:**
    - Explain the potential impact of the project on the beneficiaries and the community.
    - Describe the benefits that the project will bring and the positive changes it will create.

11. **Conclusion:**
    - Summarize the key points of the proposal.
    - Reiterate the importance of the project and the need for funding.
    - Thank the potential funder for their consideration.

12. **Attachments:**
    - Include any supporting documents, such as photos, testimonials, letters of support, or relevant research.

Remember to tailor the proposal to the specific funder's guidelines and requirements. A well-written and comprehensive proposal can significantly increase the chances of securing funding for your NGO's important initiatives.

 The Comprehensive Guide On CSR For Consultant

Annexure 4: CSR Impact Assessment Report

Introduction

An impact assessment study report in Corporate Social Responsibility (CSR) is a comprehensive document that evaluates the effectiveness and outcomes of CSR initiatives. It aims to measure the social, environmental, and economic impacts of the project and provides valuable insights for stakeholders, donors, and the organization itself. This detailed Structure outlines the key components and structure of a CSR Project Impact Assessment Study Report.

Executive Summary:

The executive summary is a concise overview of the impact assessment study report. It should provide a brief introduction to the CSR project, the purpose of the impact assessment, key findings, and the significance of the project's impacts. This section should be written in a way that captures the attention of readers and encourages them to read the entire report.

Introduction:

The introduction sets the context for the impact assessment study. It should provide background information about the CSR project, including its objectives, target beneficiaries, and the specific issues it aims to address. Describe the scope of the impact assessment and mention any previous studies or evaluations that have been conducted.

Theory of Change (ToC) or Logic Model:

Explain the Theory of Change (ToC) or logic model that guided the design and development of the CSR project. This model outlines the cause-and-effect relationships between project activities, outputs, outcomes, and impacts. It helps to understand how the project is expected to achieve its intended results.

Methodology:

Detail the research design and methodology used for the impact assessment study. This section should include information about data collection methods (surveys, interviews, focus groups, etc.), sample size, selection criteria, and any limitations or challenges faced during the study. Clearly explain how data was collected, analyzed, and interpreted.

Baseline Data:

Present the baseline data collected before the implementation of the CSR project. Baseline data serves as a reference point for comparing the project's outcomes and impacts. Include relevant demographic, social, and economic information about the target beneficiaries and the project area.

Key Findings:

   The Comprehensive Guide On CSR For Consultant

Provide a comprehensive analysis of the key findings from the impact assessment. Categorize the impacts into social, environmental, and economic dimensions. Use both qualitative and quantitative data to support the findings. This section should clearly demonstrate the changes or improvements that have occurred as a result of the CSR project.

Social Impact:

Assess the social impact of the CSR project, such as improvements in the quality of life, education, health, and social cohesion within the community. Use case studies, testimonials, and anecdotes to illustrate the positive changes and the project's contribution to the well-being of the beneficiaries.

Environmental Impact:

Evaluate the environmental impact of the CSR project, focusing on aspects such as resource conservation, waste reduction, and sustainable practices. Discuss any changes in environmental conditions resulting from the project and highlight the organization's efforts towards environmental sustainability.

Economic Impact:

Analyze the economic impact of the CSR project, including job creation, income generation, and overall economic development in the community. Quantify any direct and indirect economic benefits brought about by the project. Consider the project's contribution to the local economy and the livelihoods of the beneficiaries.

Stakeholder Engagement and Participation:

 The Comprehensive Guide On CSR For Consultant

Describe how stakeholders were engaged throughout the impact assessment study. Highlight the involvement of the beneficiaries, local community members, and other relevant stakeholders in the assessment process. Acknowledge their perspectives and inputs.

Lessons Learned:

Discuss the lessons learned during the project's implementation, including challenges faced and best practices identified. Reflect on the strengths and weaknesses of the project and provide recommendations for future improvements.

Recommendations:

Based on the impact assessment findings, present actionable recommendations for improving the current CSR project and informing future projects. These recommendations should be specific, measurable, achievable, relevant, and time-bound (SMART). Consider the sustainability and scalability of the project.

Conclusion:

Summarize the main findings and conclusions of the impact assessment study. Reiterate the significance of the project's impacts and how it aligns with the organization's overall CSR goals and objectives. Emphasize the value of the project in creating positive social and environmental change.

Annexes:

Include any additional information, charts, graphs, or tables that support the findings and analysis presented in the

 The Comprehensive Guide On CSR For Consultant

report. This may include survey questionnaires, interview transcripts, and detailed data tables.

References:

List all the sources and references used in the impact assessment study report.

In conclusion, a well-structured CSR Project Impact Assessment Study Report is an essential tool for evaluating the effectiveness of CSR initiatives and communicating the organization's social impact to stakeholders and donors. It should provide a comprehensive analysis of the project's outcomes and impacts while offering actionable recommendations for future improvements.

The Comprehensive Guide On CSR For Consultant

Annexure 5: CSR Initiatives Theory of Change

The Theory of Change (ToC) is a powerful and widely used tool in program planning, design, and evaluation. It is a systematic approach that outlines the causal pathways through which a program or intervention is expected to bring about the desired outcomes and impact. In simpler terms, it explains how and why a particular set of activities is believed to lead to specific results and eventually contribute to the overall goal of the project.

The key elements of a Theory of Change include:

Vision and Goals: It starts with a clear vision of the change the program seeks to achieve and the long-term goals that align with that vision. These goals are usually broad and represent the intended impact of the intervention.

Outcomes: Outcomes are the intermediate changes or results that need to occur to achieve the long-term goals. They represent the desired changes in behavior, knowledge, skills, attitudes, or conditions among the target population.

Causal Pathway: The ToC maps the causal pathway, illustrating the logical connections between the inputs (resources and activities), the outputs (immediate results of the activities), the outcomes (intermediate changes), and the long-term impact.

Assumptions: Along the causal pathway, there are underlying assumptions or hypotheses about how each step will lead to the desired outcomes. These assumptions need to be made explicit and validated during the program implementation and evaluation.

Indicators: Indicators are specific and measurable variables that allow the monitoring and evaluation of progress toward achieving the outcomes and goals. They help track the success of the program and assess whether it is on track to achieve its intended results.

Evaluation: A well-developed Theory of Change guides the program evaluation by providing a clear Structure for assessing the effectiveness and impact of the intervention. It helps in understanding what needs to be measured and evaluated to determine success.

The Theory of Change is not just a linear representation of cause-and-effect; it often includes feedback loops and acknowledges that multiple factors can influence the outcomes. It allows program planners and evaluators to make explicit the underlying assumptions and logic behind the program's theory of how it will bring about change.

The process of developing a Theory of Change involves engaging stakeholders, including beneficiaries, experts, and program implementers, to gain a shared understanding of the intervention's theory and to ensure that all perspectives are considered in designing and evaluating the program.

Education Project: To illustrate how the Theory of Change (ToC) can be used in social development, let's consider a real example of a literacy program aimed at improving literacy rates among children in a disadvantaged community:

Vision and Goals:
The vision is to improve literacy rates and educational outcomes among children in the community, leading to better opportunities and long-term empowerment. The specific goal is to increase the reading and writing proficiency of children aged 6 to 12 within three years.

Outcomes:
The intermediate outcomes are identified based on the program's logic and evidence from research and past experiences. For this example, the outcomes could include:
Outcome 1: Improved access to quality educational resources and materials.
Outcome 2: Enhanced teaching methods and teacher training.
Outcome 3: Increased parental involvement in their children's education.
Outcome 4: Improved reading and writing skills among children.

Causal Pathway:
The ToC maps out the logical connections between inputs, activities, outputs, and outcomes. For instance:
Input 1: Funding and resources for the literacy program.
Activity: Establishing a community learning center with a library and educational materials.
Output: Increased availability of books, educational games, and learning resources.

Outcome: Improved access to quality educational resources (Outcome 1).

Input 2: Training workshops for teachers on effective literacy instruction techniques.
Activity: Conducting regular teacher training sessions.
Output: Teachers equipped with new teaching methods and strategies.
Outcome: Enhanced teaching methods and teacher training (Outcome 2).

Assumptions:
Throughout the causal pathway, there will be underlying assumptions. For example:
Assumption: Access to quality educational resources will motivate children to engage in reading and learning.
Assumption: Teachers who receive training will effectively implement new literacy instruction methods in the classroom.
Indicators:
Measurable indicators are used to assess progress and success. Examples of indicators for the outcomes are:

Outcome 1 Indicator: Number of children using the community learning center per month.
Outcome 2 Indicator: Improvement in teacher evaluation scores after training.
Evaluation: The ToC helps guide the evaluation process by identifying what should be measured and how to assess progress towards outcomes and goals. Through regular monitoring, the program can determine whether it is on track to achieve the desired impact.
By using the Theory of Change, the literacy program can make explicit the assumptions and logic behind its approach to improving literacy rates. This clarity helps in

   The Comprehensive Guide On CSR For Consultant

the effective design and implementation of the program and allows stakeholders to understand the expected outcomes and how they will be achieved. It also enables the program to track progress, make adjustments as needed, and ultimately achieve meaningful and sustainable social development outcomes.

Annexure 6: CSR Initiative Monitoring and Evaluation

Structure of Monitoring and Evaluation (M&E) Tools for Corporate Social Responsibility (CSR) Projects:

Indicator Matrix: Develop a comprehensive list of indicators that align with the project's objectives and outcomes. The indicator matrix should include both quantitative and qualitative indicators, making it easier to track progress and measure the project's success.

Data Collection Plan: Create a detailed data collection plan that outlines the sources of data, data collection methods, data collection schedule, responsible persons, and data quality assurance procedures. The plan should specify who will collect the data, how often it will be collected, and how it will be stored and managed.

Checklists: Use checklists to monitor the completion of project activities and ensure that all necessary steps are carried out. Checklists are useful for ensuring that tasks are completed on time and in the correct sequence.

Activity Logs: Keep activity logs to record the dates, times, and details of project activities and events. Activity logs provide a chronological record of project implementation and serve as a reference for progress tracking.

Progress Reports: Regularly update progress reports to track achievements, challenges, and milestones. Progress reports should include quantitative data, qualitative insights, and visual representations of key indicators.

Quantitative Data Collection Tools:

- **Questionnaires:** Design questionnaires to collect quantitative data from beneficiaries and stakeholders, capturing their perceptions and experiences related to the project. Use closed-ended questions for standardized data collection.
- **Surveys:** Conduct surveys to gather data from a broader sample of beneficiaries or the community. Surveys can be administered in person, by phone, or online.

Qualitative Data Collection Tools:

- **Interview Guides:** Develop interview guides to conduct structured or semi-structured interviews with project stakeholders, beneficiaries, and partners. Use open-ended questions to capture rich qualitative data.
- **Focus Group Discussion (FGD) Guides:** Use FGD guides to facilitate group discussions with community members to gather qualitative insights and diverse perspectives.

 The Comprehensive Guide On CSR For Consultant

- **Beneficiary Feedback Mechanism:** Implement a feedback mechanism that allows beneficiaries to express their thoughts, suggestions, and concerns about the project's implementation and impact. Feedback forms, suggestion boxes, or community meetings can be used for this purpose.
- **GIS Mapping:** Utilize Geographic Information System (GIS) mapping to visualize project activities and their spatial distribution, enabling better planning, analysis, and decision-making.

Data Management System: Establish a data management system to organize, store, and analyze data collected during monitoring. Use spreadsheets, databases, or specialized M&E software for efficient data management.

Data Analysis Tools:

- **Statistical Software:** Use statistical software like SPSS, Excel, or R for quantitative data analysis. Statistical analysis helps in identifying trends, patterns, and correlations within the data.
- **Qualitative Analysis Software:** Utilize qualitative analysis software such as NVivo to analyze textual data from interviews and focus group discussions. Qualitative analysis provides in-depth insights into beneficiary experiences and perceptions.

Data Visualization Tools: Use data visualization tools like charts, graphs, and infographics to present monitoring data in a visually appealing and easy-to-understand manner. Visual representations help stakeholders grasp the project's progress and impact quickly.

   The Comprehensive Guide On CSR For Consultant

Learning and Reflection Mechanisms: Encourage learning and reflection within the project team through regular review meetings, lessons learned workshops, and knowledge sharing sessions. Reflecting on monitoring data helps identify areas for improvement and innovation.

Feedback and Dissemination Strategy: Develop a strategy for sharing monitoring findings with relevant stakeholders, including beneficiaries, donors, partners, and the wider public. Dissemination can be through reports, presentations, or interactive workshops.

Mid-term and End-line Evaluations: Plan for mid-term and end-line evaluations to assess the project's effectiveness, relevance, efficiency, and sustainability. These evaluations provide valuable insights for project improvement and decision-making.

Impact Assessment Tools:

- **Theory of Change (ToC):** Use the Theory of Change to map out the causal pathways between project activities and intended impacts.
- **Counterfactual Analysis:** Employ counterfactual analysis (e.g., control group comparisons) to determine the project's attribution to observed changes.

Learning and Reflection Mechanisms: Encourage learning and reflection within the project team through regular review meetings, lessons learned workshops, and knowledge sharing sessions.

Feedback and Dissemination Strategy: Develop a strategy for sharing monitoring and evaluation findings

 The Comprehensive Guide On CSR For Consultant

with relevant stakeholders, including beneficiaries, donors, partners, and the wider public.

In conclusion, a well-structured Structure of monitoring tools ensures that data is collected systematically, stakeholders are engaged, and progress is continuously tracked. These tools help project managers and stakeholders make informed decisions, address challenges, and optimize the impact of CSR projects.

Annexure 7: Employee Engagement in Corporate Social Responsibility Initiatives

Real company that has excelled in improving employee engagement in corporate social responsibility (CSR) initiatives is Salesforce. Salesforce is a leading cloud-based software company known for its commitment to social impact and sustainability. Here's how they have achieved high employee engagement in CSR initiatives:

1-1-1 Model: Salesforce has a unique 1-1-1 model, where they pledge to donate 1% of their equity, 1% of their employees' time, and 1% of their product to nonprofit organizations and charitable causes. This integrated approach to CSR ensures that employees actively participate in giving back to the community.

Volunteer Time Off (VTO) Policy: Salesforce offers employees paid time off specifically for volunteering activities. This policy allows employees to engage in social impact projects they are passionate about during work hours, encouraging them to get involved and make a difference.

Trailblazer Community: Salesforce has a vibrant "Trailblazer Community" that brings together employees, customers, partners, and stakeholders to collaborate on CSR initiatives. The platform enables employees to connect with like-minded individuals and work together towards shared social goals.

Employee Resource Groups (ERGs): Salesforce has several Employee Resource Groups focused on various social causes, such as sustainability, diversity, and social justice. These groups provide employees with opportunities to actively engage in CSR efforts aligned with their interests and values.

Earthforce: The company has an internal program called "Earthforce," which promotes environmental sustainability and empowers employees to champion green initiatives within the organization and in their communities.

Impact Labs: Salesforce has established "Impact Labs," where employees use their technical skills to develop innovative solutions for nonprofit organizations and social enterprises. This program allows employees to apply their expertise to address pressing social challenges.

Corporate Philanthropy: Salesforce actively supports nonprofits and social enterprises through corporate philanthropy. They provide grants, donations, and technical assistance to organizations working on critical social issues, inspiring employees to be part of the impact they create.

   The Comprehensive Guide On CSR For Consultant

Global Volunteer Month: Salesforce hosts an annual "Global Volunteer Month" during which employees from around the world participate in various community service projects. This concentrated effort to give back creates a sense of unity and purpose among employees.

CSR Reporting and Transparency: Salesforce publishes an annual CSR report that highlights the company's social and environmental performance. The transparent reporting allows employees to understand the impact of their CSR efforts and encourages continuous improvement.

Leadership Commitment: Salesforce's leadership, including CEO Marc Benioff, is actively involved in driving the company's CSR initiatives. Their commitment sets a strong example for employees and reinforces the importance of social responsibility throughout the organization.

Through these initiatives, Salesforce has successfully created a culture of employee engagement in CSR. Their comprehensive approach to giving back and empowering employees to make a positive impact has contributed to a strong sense of purpose and pride among their workforce.

Company: Tata Consultancy Services (TCS)

Tata Engage Program: TCS has a comprehensive corporate social responsibility program called "Tata Engage," which encourages employees to actively participate in various social and environmental initiatives. The program provides employees with opportunities to volunteer their time, skills, and expertise

---

to support communities in need.

Impact Initiatives: TCS focuses on driving initiatives that have a meaningful impact on society. They undertake projects related to education, healthcare, environmental conservation, and skill development. Employees are encouraged to be part of these initiatives, aligning their CSR efforts with their personal interests.

Pro Bono Projects: TCS offers pro bono services to nonprofit organizations and social enterprises through its "TCS Pro Bono" program. Employees can contribute their technical and business skills to help these organizations achieve their missions.

iVolunteer: TCS collaborates with iVolunteer, a platform that connects volunteers with NGOs and social projects. Through this partnership, TCS employees can easily find and participate in volunteering opportunities based on their skills and interests.

Employee Resource Groups (ERGs): TCS has established Employee Resource Groups focused on social causes, such as "TCS Cares." These groups organize volunteering events and drives, fostering a sense of community and purpose among employees engaged in CSR initiatives.

Global Village Program: TCS's "Global Village" program allows employees to work on community development projects in remote and underprivileged areas around the world. This program offers a unique opportunity for employees to experience cross-cultural engagement while making a positive impact.

     The Comprehensive Guide On CSR For Consultant

CSR Awards and Recognition: TCS recognizes and appreciates employees who actively contribute to CSR initiatives through various awards and recognition programs. This motivates employees to continue their engagement in social responsibility efforts.

Sustainable Initiatives: TCS emphasizes sustainability and green practices within the organization. By involving employees in eco-friendly initiatives and environmental conservation projects, the company fosters a sense of responsibility towards the environment.

Collaboration with NGOs: TCS collaborates with a network of NGOs to implement CSR projects effectively. Employees are actively involved in planning and executing these projects, enabling them to witness the direct impact of their efforts.

Communication and Awareness: TCS communicates its CSR initiatives and impact regularly to employees through internal communications, newsletters, and social media. This transparent communication helps build awareness and understanding of the company's CSR goals and achievements.

Through these initiatives, Tata Consultancy Services has successfully cultivated a culture of employee engagement in corporate social responsibility, making a significant difference in the lives of communities and the environment.

   The Comprehensive Guide On CSR For Consultant

Annexure 8: Developing Ethical Business Practices In CSR

Developing a Structure for ethical business practices in corporate social responsibility (CSR) and sustainability initiatives is essential to ensure that these efforts align with ethical standards and contribute to positive societal and environmental impacts. Below is a comprehensive Structure to guide the implementation of ethical practices in CSR and sustainability initiatives:

1. Ethical Policy and Governance:

Establish a clear and comprehensive ethical policy that outlines the company's commitment to ethical behavior in all CSR and sustainability initiatives.

Designate a responsible governance body to oversee the implementation of ethical practices and monitor compliance.

2. Stakeholder Engagement and Inclusivity:

Engage with stakeholders from all levels of society, including employees, customers, suppliers, local communities, NGOs, and investors.

Consider diverse perspectives and needs to ensure inclusivity and responsiveness in CSR and sustainability planning.

3. Human Rights and Labor Practices:

Ensure that all business activities respect and promote human rights, including fair labor practices, non-discrimination, and freedom of association.

Uphold labor standards, including fair wages, safe working conditions, and protection of workers' rights.

4. Environmental Responsibility:

Integrate environmental sustainability into all aspects of business operations and CSR initiatives.

Set clear goals for reducing environmental impact, conserving resources, and minimizing waste and emissions.

5. Supply Chain Ethics:

Implement ethical sourcing and supply chain practices to ensure that suppliers and partners adhere to the same ethical standards.

 The Comprehensive Guide On CSR For Consultant

Conduct regular assessments and audits to monitor and address ethical issues in the supply chain.

6. Transparency and Reporting:

Practice transparent reporting on CSR and sustainability initiatives, including their goals, progress, and outcomes.

Publicly disclose relevant information to stakeholders, promoting accountability and building trust.

7. Philanthropy and Community Investment:

Engage in philanthropic activities and community investment with a focus on addressing social needs and promoting sustainable development.

Collaborate with local communities to identify their needs and involve them in the design and implementation of community development projects.

8. Anti-Corruption and Fair Competition:

Comply with anti-corruption laws and adopt fair competition practices in all business dealings.

Foster a culture of integrity and ethical decision-making within the organization.

9. Continuous Improvement and Impact Measurement:

Continuously assess and improve the effectiveness of ethical practices in CSR and sustainability initiatives.

Implement robust impact measurement methods to evaluate the social and environmental outcomes of these initiatives.

 The Comprehensive Guide On CSR For Consultant

10. Employee Training and Ethical Culture:

Provide regular ethics training to employees to promote ethical awareness and decision-making.

Foster an ethical culture within the organization by recognizing and rewarding ethical behavior.

11. Global Standards and Initiatives:

Align CSR and sustainability efforts with global standards, such as the United Nations Sustainable Development Goals (SDGs) and Global Reporting Initiative (GRI) guidelines.

Engage in industry-specific initiatives and collaborations to drive collective progress towards ethical and sustainable practices.

12. Collaboration and Learning:

Collaborate with other companies, NGOs, and stakeholders to share best practices and learn from one another.

Participate in industry-wide initiatives to drive positive change on a broader scale.

By developing and implementing a robust Structure for ethical business practices in CSR and sustainability initiatives, companies can ensure that their efforts align with ethical principles, contribute to sustainable development, and positively impact society and the environment. This approach not only strengthens the company's reputation and brand value but also creates a lasting positive legacy for future generations.

 The Comprehensive Guide On CSR For Consultant

Annexure 9: Social Return on Investment (SROI)

Let's consider a hypothetical Social Return on Investment (SROI) example for a Corporate Social Responsibility (CSR) and Sustainability project:

Real Company: Coca-Cola

Project Name: "Replenish Africa Initiative" (RAIN)

Objective: The Coca-Cola Company launched the RAIN project in 2009 with the goal of replenishing the water used in its beverages and supporting sustainable water access, sanitation, and hygiene (WASH) initiatives in Africa. The project aimed to improve water availability and sanitation for communities, conserve water resources, and create positive social and environmental impacts.

Activities:

Water Replenishment: Coca-Cola focused on achieving water neutrality by returning to communities and nature the equivalent amount of water used in its finished beverages and their production.

Water Access and Sanitation: The company partnered with local governments, NGOs, and communities to implement clean water access and sanitation projects, such as constructing water boreholes, rainwater harvesting systems, and sanitation facilities.

Water Conservation: Coca-Cola worked to implement water conservation initiatives within its manufacturing plants, distribution centers, and communities to minimize water usage and promote efficient water management practices.

SROI Calculation:

The SROI analysis for Coca-Cola's "Replenish Africa Initiative" involved a comprehensive assessment of the project's inputs, outputs, outcomes, and impacts:

Inputs: The total investment made by Coca-Cola in terms of financial resources, staff time, and infrastructure for the RAIN initiative.

Outputs: The measurable and immediate outcomes of the project, such as the number of water projects implemented, the volume of water replenished, and the number of communities with improved water access.

 The Comprehensive Guide On CSR For Consultant

Outcomes: The short and medium-term changes resulting from the project, such as improved health and well-being of community members, enhanced economic opportunities, and increased awareness of water conservation practices.

Impact: The long-term and broader changes attributable to the project, including enhanced community resilience, increased school attendance due to improved water and sanitation facilities, and greater ecosystem sustainability.

Value the Outcomes and Impact: Assign a monetary value to the social and environmental outcomes and impacts achieved by the project. This valuation often involves assessing the benefits provided to stakeholders, local economies, and the environment.

Calculate SROI Ratio: Divide the total value of outcomes and impact by the total investment (inputs) to obtain the SROI ratio.

It's worth mentioning that SROI analysis is a dynamic process that requires continuous monitoring and evaluation to account for changes and adaptations over time. For the most current and detailed information on Coca-Cola's "Replenish Africa Initiative," including its SROI analysis, I recommend referring to official reports and publications from Coca-Cola or independent evaluations conducted by third-party organizations.

Real Company: Tata Group

Social Return on Investment (SROI) can be calculated for a different hypothetical social program:

 The Comprehensive Guide On CSR For Consultant

Program Name: ABC Food Security Initiative

Inputs:

Total program budget: INR 200,000
Staff salaries and benefits: INR 80,000
Purchase of seeds and agricultural inputs: INR 40,000
Training and capacity-building expenses: INR 15,000
Equipment and tools: INR 10,000
Other operational expenses: INR 5,000
Total input value: INR 150,000
Outputs:

Number of participating households: 100
Duration of the program: 1 year
Number of training sessions: 8
Average participant attendance rate: 90%
Total participant contact hours: 720 (100 households x 8 sessions x 90% attendance x 1 hour)
Outcomes:

Increased crop yield: Participants reported a 20% increase in crop production.
Improved household food security: Participants reduced their monthly expenditure on food by 25%.
Enhanced nutrition: Participants reported an increase in the consumption of fruits and vegetables.
Valuation:

Increased crop yield:
Value of increased crop production: Estimated at INR 5,000 per household per year
Improved household food security:
Monthly savings on food expenditure: Estimated at INR 100 per household per month

 The Comprehensive Guide On CSR For Consultant

Enhanced nutrition:
Value of improved nutrition and health outcomes:
Estimated at INR 1,000 per household per year
Calculation:

Social Value Created:

Increased crop yield: 100 households x INR 5,000 = INR 500,000
Improved household food security: 100 households x (INR 100 x 12 months) = INR 120,000
Enhanced nutrition: 100 households x INR 1,000 = INR 100,000
Total social value created: INR 720,000
SROI Ratio:

SROI Ratio = Total social value created / Total program investment
SROI Ratio = INR 720,000 / INR 150,000
SROI Ratio = 4.8
SROI Impact:

For every INR 1 invested in the ABC Food Security Initiative, a social value of INR 4.8 is created.
Again, please note that the figures used in this example are fictional and should be based on actual research, expert knowledge, or evaluation data specific to the program context. The SROI calculation provides an estimate of the social value generated by the program and helps assess the efficiency and effectiveness of the intervention.

 The Comprehensive Guide On CSR For Consultant

Annexure 10: Standard Non-Governmental Organizations (NGOs) Reporting Structure

To comply with Corporate Social Responsibility (CSR) provisions, Non-Governmental Organisations (NGOs) have developed a variety of reporting structures. However, many NGOs engage in social and developmental activities comparable to those covered by corporate social responsibility initiatives. NGOs are not required to adhere to the same reporting requirements as corporations, but they may voluntarily implement similar reporting practises to demonstrate their impact, transparency, and accountability to stakeholders and donors. Listed below is a suggested Structure for NGOs writing a comprehensive CSR-style report:

[NGO Name] - Corporate Social Responsibility Report [Year]

Introduction:

Provide a brief overview of the NGO's mission, vision, and focus areas. Mention the objectives of the CSR-style report and its alignment with the principles of transparency and accountability.

Governance and Structure:

Explain the governance structure of the NGO, including the board members, management team, and key personnel responsible for overseeing and implementing CSR initiatives.

CSR Policy and Strategy:

Detail the CSR policy and strategy of the NGO, outlining the focus areas, target beneficiaries, and expected outcomes of the initiatives.

Key CSR Areas and Initiatives:

Enumerate the major CSR areas and initiatives undertaken by the NGO during the reporting year. These may include:

1. Education and Skill Development:

Describe programs aimed at enhancing access to quality education and vocational training for marginalized communities.

- Process (In Detail)
- Input (In Detail)
- Output (In Detail)
- Outcome (In Detail)

2. Healthcare and Sanitation:

 The Comprehensive Guide On CSR For Consultant

Highlight initiatives related to healthcare services, awareness campaigns, and sanitation projects.

- Input
- Output
- Outcome

3. Livelihood and Women Empowerment:

Detail projects focused on livelihood enhancement, entrepreneurship development, and women's economic empowerment.

- Input
- Output
- Outcome

4. Environmental Sustainability:

Discuss efforts to promote environmental conservation, waste management, and renewable energy adoption.

- Input
- Output
- Outcome

5. Community Development:

Mention infrastructure development projects and initiatives targeting overall community development.

- Input
- Output
- Outcome

 The Comprehensive Guide On CSR For Consultant

Impact Assessment:

Present an assessment of the impact of the NGO's CSR initiatives during the reporting year. Use data, case studies, and testimonials to demonstrate the positive changes brought about by the projects.

Financials:

Provide a breakdown of the funds allocated to CSR initiatives, including the sources of funding, expenditure on each project, and financial transparency.

Stakeholder Engagement:

Explain the NGO's engagement with various stakeholders, such as communities, donors, volunteers, and government agencies, to gain their support and feedback.

Transparency and Accountability:

Discuss the measures taken to ensure transparency and accountability in implementing CSR initiatives. Highlight any partnerships or collaborations with other organizations.

Challenges and Learnings:

Acknowledge any challenges faced during the implementation of CSR initiatives and share the lessons learned in addressing them.

Future Plans:

Outline the NGO's future CSR plans and its vision for expanding its impact in the upcoming years.

The Comprehensive Guide On CSR For Consultant

Conclusion:

Summarize the NGO's commitment to CSR principles, its achievements in the reporting year, and express gratitude to donors, partners, volunteers, and stakeholders for their valuable support.

[Signature of Authorized Representative]

[Name and Designation]

Annexure 11: CSR Case study success stories writing

Corporate social Responsibility or Social Development Initiative

Writing compelling and effective case studies and success stories for social development initiatives requires a standard Structure that captures the essence of the project, its impact, and the beneficiaries' experiences. Below is a standard Structure to follow when writing social development initiative case studies and success stories:

Introduction:

Provide a brief introduction to the social development initiative, including its objectives, target beneficiaries, and the organization or company responsible for implementing it.

Background and Context:

 The Comprehensive Guide On CSR For Consultant

Describe the background and context of the social development issue the initiative aims to address.

Include relevant data, statistics, or facts to provide context and highlight the significance of the project.

Problem Statement:

Clearly state the problem or challenge the social development initiative seeks to solve.

Explain why the problem is critical and the potential consequences if left unaddressed.

Approach and Implementation:

Detail the approach taken to implement the social development initiative.

Describe the strategies, methodologies, and activities employed to achieve the project's objectives.

Mention any partnerships or collaborations that contributed to the initiative's success.

Key Milestones and Achievements:

Highlight the key milestones achieved during the implementation of the initiative.

Provide quantitative data and qualitative anecdotes that demonstrate the initiative's progress and success.

Impact and Outcomes:

Describe the impact of the social development initiative on the target beneficiaries and the community as a whole.

Present both qualitative and quantitative data to showcase the outcomes and changes brought about by the project.

Include testimonials, quotes, or stories from beneficiaries to add a human element to the impact assessment.

Challenges and Solutions:

Discuss any challenges faced during the implementation of the social development initiative.

Explain how these challenges were overcome and the solutions employed to address them.

Sustainability and Scalability:

Address the sustainability of the initiative and its potential for scalability.

Discuss how the project's impact can be sustained over time and expanded to reach more beneficiaries.

Lessons Learned:

Share the key lessons learned from the social development initiative.

Discuss insights gained from the project's implementation that could be helpful for similar initiatives in the future.

Conclusion:

Summarize the overall success and impact of the social development initiative.

Emphasize the importance of continued efforts in social development and the role of the initiative as a model for positive change.

 The Comprehensive Guide On CSR For Consultant

Visuals and Media:

Include relevant photographs, infographics, or videos to support and enhance the storytelling.

Visuals can help bring the case study to life and make it more engaging for readers.

By following this standard Structure, you can create well-structured and compelling case studies and success stories that effectively communicate the impact and success of your social development initiatives. These stories can serve as valuable tools for raising awareness, attracting support from stakeholders, and inspiring others to take part in similar meaningful initiatives.

Structure for Individual beneficiary's case study and success stories

Writing case studies and success stories focusing on individual beneficiaries of social development initiatives requires a thoughtful and sensitive approach to capture their experiences, challenges, and the impact of the project on their lives. Below is a standard Structure to guide you in writing individual beneficiary case studies and success stories:

Introduction:

Start with a brief introduction, including the name and background of the beneficiary.

Mention the social development initiative they were part of and its objectives.

Background and Context:

 The Comprehensive Guide On CSR For Consultant

Provide some background information about the beneficiary's life before the initiative.

Describe the social or economic challenges they faced, highlighting the problem the initiative aimed to address.

Journey and Participation:

Narrate the beneficiary's journey through the initiative, from their initial involvement to their level of participation and engagement.

Explain how they became connected to the initiative and what motivated them to participate.

Challenges and Aspirations:

Discuss the specific challenges and obstacles the beneficiary encountered in their life.

Include their aspirations, hopes, and dreams for the future.

Initiative's Impact:

Describe the ways in which the social development initiative impacted the beneficiary's life.

Include both tangible and intangible changes, such as improved access to education, better healthcare, increased income, enhanced skills, or increased confidence and self-esteem.

Personal Growth and Empowerment:

Highlight the personal growth and empowerment experienced by the beneficiary as a result of the initiative.

   The Comprehensive Guide On CSR For Consultant

Share stories of how they overcame challenges, developed new skills, and gained confidence.

Testimonials and Quotes:

Include direct quotes or testimonials from the beneficiary, if possible.

Their own words can add authenticity and emotional resonance to the case study.

Visuals:

Include photographs or visuals that showcase the beneficiary's journey and the impact of the initiative on their life.

Use images to evoke emotions and provide a visual narrative.

Future Outlook:

Discuss the beneficiary's future outlook and how the changes brought about by the initiative have influenced their aspirations.

Mention any long-term plans or goals they now have.

Conclusion:

Summarize the impact of the social development initiative on the individual beneficiary's life.

Emphasize the significance of such initiatives in bringing positive change to the lives of individuals and communities.

Privacy and Consent:

Ensure that you have obtained the necessary consent and permissions from the beneficiary before sharing their personal information and stories.

Respect their privacy and confidentiality throughout the case study.

Remember that individual beneficiary case studies and success stories are about people and their personal journeys. Use empathetic language, focus on their experiences, and highlight the transformative power of social development initiatives in their lives. Additionally, ensure that the stories are accurate and respectful, maintaining the dignity of the beneficiaries portrayed.

Annexure 12: Social Development Project Management Cycle

The phases of a social development project management cycle are similar to those of a typical project management cycle but tailored specifically for projects that aim to address social issues or bring about positive social change. let's go through the phases of a social development project management cycle with an example of a project aimed at improving access to education for marginalized children in a rural community:

Initiation:

In this phase, the social development project is conceptualized, and the need for the project is identified. The project's goals, objectives, and scope are defined, and initial stakeholders are identified.

 The Comprehensive Guide On CSR For Consultant

Example: A nonprofit organization identifies that many children in a rural community are not attending school due to lack of access and resources.

Activities: The organization conducts a needs assessment to understand the specific challenges and assess the community's willingness to support the project.

Planning:

Detailed planning takes place in this phase. A project plan is created, outlining the project's activities, resources required, timeline, budget, and risks. Stakeholder engagement strategies are developed, and roles and responsibilities are assigned.

Example: The organization plans the project and sets the goal to establish a community school to provide education to the marginalized children.

Activities: The team creates a detailed project plan that includes budget allocation, resource requirements, timeline, and identifies key stakeholders, such as local government officials, parents, and community leaders.

Design and Development:

During this phase, the project design is finalized, and project activities and interventions are developed. Social impact assessments may be conducted to understand potential consequences, both positive and negative.

Example: The organization designs the community school, taking into account the specific needs and cultural aspects of the community.

    The Comprehensive Guide On CSR For Consultant

Activities: The team develops the curriculum, recruits and trains local teachers, and secures necessary resources like school supplies and infrastructure.

Implementation:

This is the phase where the project is put into action. The project team carries out the planned activities, and resources are allocated accordingly. Stakeholder engagement and communication are critical during this phase.

Example: The community school is built and starts functioning, offering education to marginalized children.

Activities: The teachers conduct classes, and the organization monitors the implementation, ensuring the school is running smoothly and effectively.

Monitoring and Evaluation:

Throughout the project's implementation, progress is continually monitored against the set objectives and performance indicators. Evaluation processes help assess the project's effectiveness and identify areas for improvement.

Example: The organization closely monitors the school's progress and its impact on children's education and well-being.

Activities: Regular assessments of students' academic performance and overall development are conducted. Feedback from teachers, parents, and students is gathered.

Documentation and Reporting:

 The Comprehensive Guide On CSR For Consultant

All aspects of the project, including activities, outcomes, challenges, and lessons learned, are documented. Regular reports are generated and shared with stakeholders, donors, and partners.

Example: The organization documents the progress, challenges, and outcomes of the community school project.

Activities: Reports are generated at specific intervals to share with donors, stakeholders, and the broader community, showcasing the project's achievements and the improvements in children's access to education.

Scale-up and Replication:

If the project is successful and impactful, there might be efforts to scale up its reach or replicate it in other areas to further its positive influence.

Example: The successful implementation of the community school inspires other communities facing similar challenges to adopt a similar approach.

Activities: The organization supports the replication of the model in other rural areas, collaborating with local partners and stakeholders.

Sustainability and Exit Strategy:

Social development projects should be designed with sustainability in mind. In this phase, plans are made for the project's continuation or transition after the funding period ends. Local community involvement and capacity building are vital for long-term sustainability.

Example: The organization ensures the long-term sustainability of the community school.

 The Comprehensive Guide On CSR For Consultant

Activities: The community is empowered to take ownership of the school's management and find sustainable funding sources, reducing dependence on external support.

Learning and Adaptation:

Lessons learned from the project's successes and failures are analyzed to improve future projects. Adaptations may be made based on these insights.

 Example: The organization learns from the project's experiences to improve future projects.

Activities: Lessons learned from the project help refine the approach for similar initiatives and guide decision-making in future social development projects.

Closure:

This is the final phase of the project management cycle. The project is formally closed, and a comprehensive report is prepared. A final review is conducted to assess the overall impact and achievements of the project.

Example: After several years of successful operation, the organization formally closes the project.

Activities: Final evaluations are conducted to assess the overall impact of the project on education outcomes and the community's well-being.

Remember that each social development project is unique, and the activities and phases may vary based on the specific context and objectives. The key is to have a well-structured and adaptable project management cycle that aligns with the needs and goals of the community being served.

 The Comprehensive Guide On CSR For Consultant

Throughout the entire social development project management cycle, it's crucial to maintain strong stakeholder engagement, adhere to ethical guidelines, and be responsive to the needs and aspirations of the communities being served. Social development projects should also be conducted with cultural sensitivity and respect for local customs and traditions.

Annexure 13: Stakeholder Engagement Strategy For CSR Initiatives

Developing a robust stakeholder engagement strategy is crucial for the success of social development initiatives. Effective engagement ensures that the concerns, needs, and perspectives of all relevant stakeholders are considered, fostering collaboration and increasing the chances of achieving positive outcomes. Here's a step-by-step guide to creating a stakeholder engagement strategy for social development initiatives:

Identify and prioritize stakeholders:

Begin by identifying all stakeholders involved in or affected by the social development initiative. Stakeholders

 The Comprehensive Guide On CSR For Consultant

may include community members, government officials, NGOs, local businesses, academic institutions, and beneficiaries. Prioritize them based on their level of influence and impact on the project.

Example

Community members (parents, children, elders)

Local government officials

Non-governmental organizations (NGOs) working in the area

Local businesses and philanthropists

Teachers and educators

Academic institutions

Understand stakeholders' interests and expectations:

Conduct thorough research and engage in dialogue with stakeholders to understand their interests, needs, and expectations regarding the initiative. This information will help tailor the engagement approach to each stakeholder group.

Example

Conduct interviews, focus groups, and surveys to gather input from community members about their educational needs and preferences.

Engage with local government officials to understand existing policies and funding opportunities for education.

   The Comprehensive Guide On CSR For Consultant

Collaborate with NGOs to identify potential partnerships and resources.

Discuss with teachers and educators to gain insights into the curriculum and teaching methods that would be most effective.

Establish clear objectives:

Define the specific goals and outcomes you aim to achieve through stakeholder engagement. These objectives should align with the broader goals of the social development initiative.

Example:

To establish a community learning center with access to quality educational resources.

To provide educational programs for both children and adults, promoting lifelong learning.

To increase enrollment and attendance in schools through awareness campaigns.

Develop a communication plan:

Create a communication plan that outlines how you will engage with different stakeholder groups throughout the project. This plan should include communication channels (e.g., meetings, workshops, surveys, online platforms) and the frequency and format of interactions.

Example

Hold community meetings to share the initiative's goals and gather feedback.

 The Comprehensive Guide On CSR For Consultant

Organize workshops and training sessions for community members and educators.

Use social media, flyers, and local radio to disseminate information about the initiative.

Create an online platform for ongoing communication and updates.

Ensure inclusivity and diversity:

Strive for inclusivity by involving a diverse range of stakeholders, including marginalized or underrepresented groups. Consider language barriers, accessibility needs, and cultural sensitivities to ensure everyone's voices are heard.

Example

Involve representatives from different age groups, genders, and social backgrounds in planning and decision-making processes.

Translate communication materials into local languages to reach broader audiences.

Ensure that the learning center's design and curriculum cater to the specific needs of the community.

Build trust and transparency:

Establish trust with stakeholders by being transparent about the initiative's objectives, progress, challenges, and decision-making processes. Honest and open communication fosters credibility and long-term engagement.

Example

     The Comprehensive Guide On CSR For Consultant

Provide regular updates on the progress of the learning center's construction and programming.

Hold transparent discussions about funding sources and how resources are allocated.

Address concerns and feedback openly and honestly.

Involve stakeholders in decision-making:

Engage stakeholders in meaningful ways by involving them in decision-making processes. Seek their input and feedback when developing project plans, strategies, and key milestones.

Example

Form a community committee to participate in decision-making about the learning center's design and activities.

Collaborate with teachers and educators to design the curriculum and learning materials.

Provide opportunities for capacity building:

Offer capacity-building workshops and training sessions for stakeholders, particularly community members and local organizations. This empowers them to actively participate and contribute to the initiative's success.

Example

Organize training sessions for local teachers and volunteers to enhance their teaching skills.

Offer adult education programs to improve literacy and life skills for community members.

 The Comprehensive Guide On CSR For Consultant

Address conflicts and concerns:

Anticipate and address conflicts and concerns among stakeholders promptly. Establish a mechanism for conflict resolution and provide a safe space for stakeholders to express their grievances.

Establish a grievance mechanism for stakeholders to raise concerns and resolve conflicts.

Encourage open dialogue between different stakeholder groups to promote understanding.

Monitor and evaluate engagement effectiveness:

Example:

Regularly assess the effectiveness of the stakeholder engagement strategy. Collect feedback, measure the impact of stakeholder input, and adjust the strategy as needed to improve outcomes.

Collect feedback through surveys and evaluations to assess the impact of the initiative.

Measure the increase in school enrollment and attendance rates over time.

Track the involvement and satisfaction of stakeholders in the learning center's activities.

Celebrate successes and communicate progress:

Acknowledge and celebrate achievements and milestones with stakeholders. Regularly communicate project progress to maintain interest and engagement.

Example:

 The Comprehensive Guide On CSR For Consultant

Organize community events to celebrate achievements and milestones of the initiative.

Use success stories and testimonials from beneficiaries to inspire further engagement.

Sustain engagement beyond project completion:

Plan for post-project engagement to ensure continuity and sustainability of efforts. Consider how stakeholders can continue to collaborate and support ongoing social development initiatives.

Establish a long-term management and funding plan for the community learning center.

Foster ongoing partnerships with NGOs, academic institutions, and local businesses for continued support.

By following this stakeholder engagement strategy, the social development initiative can effectively involve all relevant parties, address community needs, and create a sustainable impact on education in the rural community.

 The Comprehensive Guide On CSR For Consultant

Annexure 14: Stakeholder Engagement Strategy For Fund Raising For CSR Initiatives

Developing a strong stakeholder engagement strategy is essential for successful fundraising for corporate social responsibility (CSR) initiatives. Engaging with stakeholders effectively can lead to increased support, credibility, and long-term partnerships. Here's a step-by-step guide for

creating a stakeholder engagement strategy for fundraising for CSR initiatives:

Stakeholder Engagement Strategy for Fundraising for a Corporate Social Responsibility (CSR) Initiative – hypothetical Example: "Project Green Horizon"

Identify key stakeholders: Identify the stakeholders relevant to your CSR initiative and fundraising efforts. These may include:

Employees: They can be powerful advocates and donors.

Customers: Engaging customers can lead to increased brand loyalty and support.

Investors/shareholders: Demonstrating a commitment to CSR can attract socially responsible investors.

Local communities: Engaging with local communities affected by the CSR initiative can build trust and support.

NGOs and other organizations: Seek partnerships with NGOs and other organizations aligned with your CSR goals.

Government and regulatory bodies: Understand the regulations and incentives related to CSR initiatives.

Example

Employees: Engage employees as donors and volunteers for CSR activities.

Customers: Seek donations from loyal customers and inform them about the impact of their contributions.

Investors/shareholders: Highlight the CSR initiative's positive effect on the company's long-term sustainability.

   The Comprehensive Guide On CSR For Consultant

Local communities: Involve community members in planning and implementing the CSR projects.

NGOs and other organizations: Collaborate with environmental NGOs to extend the reach and impact of the initiative.

Government and regulatory bodies: Stay informed about relevant regulations and incentives for CSR activities.

Understand stakeholder interests and motivations: Research and analyze the interests, concerns, and motivations of each stakeholder group regarding your CSR initiative. Tailor your fundraising messaging to align with their values and priorities.

Example

Employees: Emphasize the company's commitment to sustainability and employee well-being.

Customers: Showcase the tangible benefits of the CSR initiative on the environment and local communities.

Investors/shareholders: Demonstrate how CSR efforts contribute to the company's brand reputation and social impact.

Local communities: Address specific community needs and involve them in decision-making.

NGOs and other organizations: Highlight opportunities for meaningful partnerships to amplify CSR efforts.

Government and regulatory bodies: Ensure compliance with regulations and explore potential incentives.

 The Comprehensive Guide On CSR For Consultant

Define clear fundraising objectives: Set specific and measurable fundraising objectives for your CSR initiative. Determine the amount of funds needed and the timeline for fundraising.

Example:

Raise INR 500,000 within the next six months for Project Green Horizon.

Secure at least 10 corporate partnerships to co-fund specific sustainability projects.

Develop a targeted communication plan: Create a communication plan that includes various channels to engage with stakeholders. Use a mix of methods such as:

Personalized emails and letters to key stakeholders.

Social media campaigns to reach a broader audience.

Events and webinars to share the CSR initiative's impact and progress.

Collaborations with local media outlets to amplify your message.

Example

Create a microsite dedicated to Project Green Horizon, showcasing its goals and impact.

Utilize social media campaigns to reach a broad audience and promote donation drives.

Host virtual webinars featuring environmental experts and beneficiaries.

 The Comprehensive Guide On CSR For Consultant

Organize community events where stakeholders can learn about the initiative.

Showcase the impact of the CSR initiative: Share success stories and real-life examples of how the CSR initiative has positively impacted communities or the environment. Use data and metrics to demonstrate the effectiveness of previous initiatives.

Example:

Highlight past CSR successes and the positive outcomes they have achieved.

Share testimonials from community members and employees impacted by previous initiatives.

Provide data on environmental improvements resulting from CSR efforts.

Leverage employee engagement: Involve employees in the fundraising process by organizing employee giving programs or encouraging volunteerism. Engaged employees can be enthusiastic advocates for your CSR initiatives.

Example

Encourage employees to participate in volunteer activities and fundraising events.

Establish a company matching gift program to double employee contributions.

Recognize outstanding employee volunteers through internal communications.

   The Comprehensive Guide On CSR For Consultant

Engage in corporate partnerships: Seek partnerships with other corporations or businesses that share similar CSR values. Co-fundraising or co-sponsoring initiatives can extend the reach of your efforts.

Example:

Identify companies with aligned CSR goals and propose co-funding opportunities.

Collaborate with local businesses to sponsor community events.

Create joint marketing materials to promote the initiative and its corporate partners.

Foster long-term relationships: Focus on building lasting relationships with stakeholders, not just one-time fundraising efforts. Continuous engagement can lead to ongoing support and contributions.

Example:

Maintain ongoing communication with stakeholders through newsletters and updates.

Organize an annual CSR summit to foster collaboration and discuss new initiatives.

Celebrate milestones and recognize stakeholders' contributions regularly.

Offer recognition and incentives: Recognize and appreciate the contributions of stakeholders through public

 The Comprehensive Guide On CSR For Consultant

acknowledgment, certificates, or awards. Consider offering incentives, such as exclusive CSR impact reports or events for major donors.

Example

Feature major donors on the Project Green Horizon microsite.

Host an annual recognition event for key donors and partners.

Offer exclusive tours of project sites to major contributors.

Transparency and accountability: Be transparent about how funds are used for CSR initiatives. Provide regular updates on the progress and impact of the initiatives to build trust and accountability.

Example

Publish regular progress reports on the microsite.

Hold quarterly webinars to provide updates on project implementation.

Respond to donor inquiries promptly and transparently.

Compliance and reporting: Ensure compliance with relevant regulations and reporting requirements for CSR initiatives. This includes adhering to tax regulations for charitable contributions.

Example

Comply with all relevant regulations regarding fundraising and CSR activities.

Submit required reports to government agencies and regulatory bodies.

Measure and evaluate fundraising efforts: Regularly assess the effectiveness of your fundraising strategy. Analyze the funds raised, donor retention rates, and the success of different communication channels.

Example

Track funds raised, donor retention rates, and engagement metrics.

Conduct surveys to gather feedback from stakeholders and improve future campaigns.

Example Outcome: By implementing this stakeholder engagement strategy for Project Green Horizon, the company successfully raised INR 600,000 within six months, surpassing its fundraising target. Ten corporate partnerships were formed, enhancing the initiative's scope and impact. Employee engagement increased significantly, with 60% of the workforce participating in volunteer activities. The CSR initiative gained positive media coverage, enhancing the company's reputation and attracting socially responsible investors. Over time, Project Green Horizon led to measurable improvements in local communities and environmental sustainability, further inspiring stakeholder commitment to the company's future CSR initiatives.

By implementing this stakeholder engagement strategy, your organization can effectively raise funds for its CSR initiatives, while also fostering meaningful relationships with stakeholders committed to making a positive social and environmental impact.

Annexure 15: Registration Of NGO Under Indian Society Registration Act

The process of registering an NGO (Non-Governmental Organization) under the Indian Society Registration Act, 1860 is as follows:

Eligibility: The organization seeking registration must have a minimum of seven members, and for certain states, the minimum number of members may be higher. The members should be Indian citizens and not be disqualified under the Act.

Name Selection: Choose a unique name for your NGO. Ensure that the name does not resemble the name of an existing organization or violate any trademarks.

Governing Body: Formulate a governing body or a managing committee. This body will be responsible for the overall management and functioning of the NGO.

Memorandum of Association (MOA): Prepare a Memorandum of Association that outlines the objectives, activities, and rules and regulations of the organization. It should be signed by all the members.

Articles of Association (AOA): Draft the Articles of Association, which define the internal regulations and procedures for the organization's functioning.

Registration Application: Complete the application for registration. It should include the following documents:

- Cover letter requesting registration
- Form of Memorandum of Association
- Form of Articles of Association
- Address proof of the registered office
- Declaration by the members/former members of the managing committee stating their willingness to be a part of the organization
- Consent letters of all members of the governing body

Any other documents required by the specific state where you are registering (if applicable)

 The Comprehensive Guide On CSR For Consultant

Submission: Submit the registration application to the Registrar of Societies or the concerned authority in your state. The application must be signed by all members of the governing body.

Verification and Approval: The authorities will verify the application and documents. If everything is in order, they will issue a Certificate of Registration under the Society Registration Act, 1860.

Certificate and PAN: Once the registration is approved, you will receive the Certificate of Registration, and you can then apply for a PAN (Permanent Account Number) for the NGO.

Please note that the process and requirements may vary slightly depending on the state in which you are registering the NGO. It's always advisable to consult with a legal professional or seek guidance from the concerned department in your state for accurate and up-to-date information.

Annexure 16: Registration of NGO under Indian Trust Registration Act

The process of registering an NGO as a trust in India is governed by the Indian Trusts Act, 1882. This act provides guidelines and regulations for the registration and administration of public trusts.

Here are the steps to register an NGO as a trust under the Indian Trusts Act, 1882:

 The Comprehensive Guide On CSR For Consultant

Eligibility: The organization should have a clear purpose or objective that falls within the permissible purposes for public trusts, such as education, healthcare, relief of poverty, advancement of religion, etc.

Name Selection: Choose a unique name for your trust. Ensure that the name is not similar to an existing trust or violates any trademarks.

Settler and Trustees: Identify the settler of the trust who will transfer the property or assets for the trust's objectives. Appoint a minimum of two trustees (beneficiaries can also be trustees) to manage the trust's affairs.

Trust Deed: Prepare a trust deed on non-judicial stamp paper, which should include the following details:

- Name and address of the trust
- Objectives and purposes of the trust
- Name and addresses of the settler and trustees
- Rules and regulations for the administration of the trust
- Details of the property or assets being transferred to the trust by the settler
- The mode of succession or appointment of new trustees
- Any other relevant clauses

Registration: Once the trust deed is prepared, it needs to be registered with the relevant Sub-Registrar or Deputy Registrar under whose jurisdiction the trust's registered office is located. Both the settler and trustees must sign the trust deed in the presence of witnesses.

The Comprehensive Guide On CSR For Consultant

Documents Required: Along with the trust deed, you will need to submit the following documents:

- Identity and address proofs of the settler and trustees
- Passport-sized photographs of the settler and trustees
- Proof of ownership of the property being transferred to the trust (if applicable)
- Any other documents as required by the concerned authority

Registration Fees: Pay the applicable registration fees at the time of submission.

Verification and Approval: The Sub-Registrar or Deputy Registrar will verify the documents and, if everything is in order, will register the trust and issue a Certificate of Registration.

Please note that the process and requirements may vary slightly depending on the state in which you are registering the trust. It's always advisable to consult with a legal professional or seek guidance from the concerned authority in your state for accurate and up-to-date information.

 The Comprehensive Guide On CSR For Consultant

Annexure 17: Registration Of Section 8 Company Under Companies Act

A Section 8 Company is a type of non-profit organization incorporated under the Companies Act, 2013 (and previously under the Companies Act, 1956). The primary objective of a Section 8 Company is to promote charitable, educational, scientific, religious, or social welfare activities without the intent of making profits. These companies are also exempt from using the words "Limited" or "Private Limited" in their names.

Here are the steps to register a Section 8 Company under the Companies Act, 2013:

Company Name Approval: Choose a unique name for the company and apply for name availability through the Ministry of Corporate Affairs (MCA) portal. Ensure that the proposed name aligns with the objectives of the company and complies with naming guidelines.

Memorandum and Articles of Association: Prepare the Memorandum of Association (MOA) and Articles of Association (AOA) of the company. The MOA should specify the charitable objectives, and the AOA should contain the rules and regulations for the company's internal functioning.

Obtain Digital Signature Certificate (DSC): The proposed directors and subscribers of the company must obtain Digital Signature Certificates (DSC) from certifying authorities. The DSC is required for filing electronic forms with the MCA.

Apply for Director Identification Number (DIN): The proposed directors must apply for Director Identification Numbers (DIN) through the MCA portal.

Incorporation Application: Prepare the incorporation application in Form SPICe (Simplified Proforma for Incorporating Company Electronically). This form includes details of the company, directors, and subscribers, along with the MOA and AOA.

Declaration and Affidavits: The proposed directors and subscribers must submit declarations and affidavits verifying the authenticity of the documents and compliance with the requirements.

Approval and Certificate: Submit the incorporation application, along with the necessary fees, to the Registrar of Companies (ROC). If the ROC is satisfied with the application and documents, they will issue a Certificate of Incorporation. In case of any discrepancies, the ROC may request clarifications or corrections.

12A and 80G Registration: After incorporation, the Section 8 Company can apply for 12A and 80G registrations under the Income Tax Act to avail tax exemptions for its donors and for itself.

Please note that the process of registering a Section 8 Company may involve several other compliance requirements, and the specific steps and documents may vary based on the individual case and the state in which the company is being registered. It's advisable to seek professional assistance from a company secretary or a legal expert to ensure a smooth and accurate registration process.

Annexure 18: Project Gantt Chart

A Gantt chart is a popular project management tool that visually represents the schedule of tasks and activities involved in a project. It helps to plan, track, and manage project progress effectively. Below is an example of a Gantt chart for a social development project. Please note that the duration and specific tasks may vary depending on the nature and scope of the project.

 The Comprehensive Guide On CSR For Consultant

Assumptions:

Project Start Date: July 1, 2023

Project End Date: December 31, 2023

Project Duration: 6 months

Tasks:

Project Initiation:

- Identify project objectives and goals
- Form project team
- Prepare project charter

Needs Assessment:

- Conduct community needs assessment
- Identify target beneficiaries
- Analyze data and finalize assessment report

Project Planning:

- Develop detailed project plan
- Allocate resources and budget
- Define project deliverables

Stakeholder Engagement:

- Identify key stakeholders
- Conduct stakeholder meetings
- Gain support and collaboration

Fundraising:

- Research potential funding sources
- Prepare grant proposals

   The Comprehensive Guide On CSR For Consultant

- Secure funding for the project

Program Development:

- Design social development programs
- Determine program components
- Create program materials and resources

Hiring and Training:

- Recruit project staff and volunteers
- Conduct training sessions
- Ensure readiness of the team

Implementation Phase 1:

- Launch social development programs
- Monitor and evaluate program effectiveness
- Address any initial challenges

Awareness Campaign:

- Develop and implement awareness campaigns
- Engage with the community through workshops, events, etc.
- Measure the impact of awareness efforts

Data Collection and Analysis:

- Gather data on project outcomes
- Analyze the data for program evaluation
- Identify areas of improvement

Implementation Phase 2:

- Continue and expand social development programs
- Incorporate lessons learned from Phase 1

 The Comprehensive Guide On CSR For Consultant

- Monitor and report on progress

Project Evaluation:

- Conduct final project evaluation
- Measure the project's overall impact
- Document successes and challenges

Project Closure:

- Prepare final project report
- Disseminate project results
- Celebrate project achievements

Note: The tasks mentioned above are just examples and may not cover all the aspects of a social development project. Additionally, the Gantt chart can be updated and revised as the project progresses and as new information or changes arise.

When creating a Gantt chart, use project management software or tools that allow you to input the task duration, dependencies, and resources to visualize the project timeline effectively. The Gantt chart will help you stay on track, manage resources efficiently, and ensure the successful completion of your social development project.

Month/ Activity	Jan	Feb	March	April	May
Activity 1	40%	30%	20%		
Activity 2	10%	40%	20%	30%	

 The Comprehensive Guide On CSR For Consultant

Activity 3		20%	40%	40%	
Activity 4			10%	30%	60%
Activity 5				60%	40%
Activity 6				10%	90%

Annexure 19: National And International CSR Awards And Applicability

Some of the National CSR Awards in India are as follows. Please note that the eligibility criteria and applicability of these awards may vary, and it's essential to refer to the respective award organizers' websites for the most current information.

National CSR Awards by the Ministry of Corporate Affairs: These awards are presented by the Indian government to recognize companies' exemplary contributions to Corporate Social Responsibility (CSR) activities. The applicability is open to companies that are required to spend on CSR activities as per the Companies Act, 2013.

FICCI CSR Awards by the Federation of Indian Chambers of Commerce and Industry (FICCI): These awards aim to recognize and encourage companies that have undertaken impactful CSR initiatives. Applicability is typically open to FICCI member companies and non-member companies operating in India.

CII-ITC Sustainability Awards by the Confederation of Indian Industry (CII) and ITC: While primarily focused on sustainability practices, these awards also acknowledge companies for their commitment to CSR initiatives. Applicability is generally open to companies across various sectors in India.

NASSCOM CSR Leadership Award by the National Association of Software and Services Companies (NASSCOM): This award recognizes the IT and software industry's contributions to social development through CSR initiatives. Applicability is usually open to NASSCOM member companies.

Golden Peacock Awards for Corporate Social Responsibility (CSR): Presented by the Institute of Directors (IOD), these awards honor organizations that have demonstrated excellence in CSR practices.

   The Comprehensive Guide On CSR For Consultant

Applicability is open to companies from various sectors across India.

TERI Corporate Environmental Awards: While primarily focused on environmental sustainability, these awards also recognize companies for their CSR efforts and impact. Applicability is generally open to companies that have undertaken significant environmental and social initiatives.

World CSR Day Awards: These awards celebrate organizations and leaders for their outstanding commitment to CSR and sustainability practices. Applicability is typically open to companies, NGOs, and individuals from various sectors.

CSR Health Impact Awards: These awards recognize companies' contributions to healthcare-related CSR initiatives. Applicability is generally open to companies involved in healthcare and related sectors.

Sustainable Plus Platinum Award: This award recognizes companies for their exceptional sustainable practices and contributions to society. Applicability is open to companies that have demonstrated sustainable and socially responsible practices.

Corporate Social Responsibility Foundation (CSRF) CSR Awards: These awards recognize organizations and individuals for their exemplary contributions to CSR activities. Applicability is typically open to companies and individuals from various sectors.

India Today Social Impact Awards: Presented by India Today Group, these awards celebrate companies, NGOs, and individuals for their significant impact on society through CSR initiatives. Applicability is generally open to companies and organizations across various sectors.

Outlook Poshan Awards: These awards, organized by Outlook India, acknowledge companies and individuals for their outstanding efforts in addressing malnutrition and promoting nutrition-related CSR initiatives. Applicability may be open to companies, NGOs, and individuals working in the nutrition and healthcare sectors.

CNBC-TV18 India Business Leader Awards (IBLA) - Responsible Corporate Leader: As part of the IBLA, this

award category recognizes business leaders who have demonstrated exemplary commitment to responsible and sustainable business practices, including CSR. Applicability is typically open to prominent business leaders from various industries.

Dun & Bradstreet Corporate Awards: These awards, organized by Dun & Bradstreet, honor companies for their overall excellence, including CSR and sustainability efforts. Applicability is generally open to companies from various sectors.

Business world Magna Awards: These awards celebrate companies and individuals for their contributions to various aspects of business, including CSR and sustainable practices. Applicability is generally open to companies and individuals from different industries.

IMC Ramkrishna Bajaj National Quality Award: While primarily focused on recognizing excellence in overall quality management, this award also considers CSR

 The Comprehensive Guide On CSR For Consultant

initiatives as a significant aspect of an organization's excellence. Applicability is open to businesses and organizations from various sectors.

Mahindra CSR Awards: Organized by the Mahindra Group, these awards acknowledge companies and NGOs for their innovative and impactful CSR projects. Applicability may be open to businesses and NGOs across different sectors.

Asia Responsible Entrepreneurship Awards (AREA): While these awards encompass responsible entrepreneurship in general, they also acknowledge companies for their outstanding CSR and sustainability practices. Applicability is generally open to businesses from various Asian countries, including India.

Global CSR Excellence & Leadership Awards: These awards recognize companies for their commitment to CSR practices and their positive impact on society and the environment. Applicability may be open to companies from India and other countries.

BSE CSR Index: While not a traditional award, the BSE (Bombay Stock Exchange) CSR Index recognizes and lists companies based on their CSR performance. It includes companies that have demonstrated strong commitment and effective implementation of CSR initiatives.

Global CSR Excellence Awards: These awards, organized by the World CSR Day, recognize companies and individuals for their outstanding contributions to CSR and sustainable practices on a global level.

 The Comprehensive Guide On CSR For Consultant

Asia Sustainability Reporting Awards: Although primarily focused on sustainability reporting, these awards also consider the CSR initiatives of companies and organizations across Asia, including India.

India CSR Awards: Presented by India CSR, these awards celebrate companies and individuals for their exceptional CSR efforts and positive impact on society.

YES BANK Natural Capital Awards: These awards focus on recognizing companies and individuals contributing to the preservation and conservation of natural capital, aligning with the principles of CSR and sustainability.

NASSCOM Foundation CSR Leadership Awards: Organized by NASSCOM Foundation, these awards acknowledge IT and software companies for their exemplary CSR contributions.

ASSOCHAM Social Banking & CSR Awards: These awards, organized by the Associated Chambers of Commerce and Industry of India (ASSOCHAM), celebrate companies and banks for their significant CSR efforts and social impact.

CSR Times Awards: Presented by CSR Times, these awards honor companies, NGOs, and individuals for their notable CSR projects and initiatives.

ET NOW - CSR Impact Awards: Organized by ET NOW, these awards recognize companies and individuals for their CSR contributions, focusing on their social impact and community development.

MPGCG CSR Awards: These awards, organized by the Madhya Pradesh Golden Chamber of Commerce and Industry, celebrate CSR initiatives and sustainable practices in the state of Madhya Pradesh.

Please note that the eligibility criteria, application process, and deadlines for these awards may change over time. To apply for any of these awards, it's important to check the specific requirements and guidelines on the respective award organizers' official websites and follow the application instructions provided.

Annexure 20: International Corporate Social Responsibility Awards

There are several prestigious international CSR awards that recognize companies and organizations for their

 The Comprehensive Guide On CSR For Consultant

outstanding commitment to Corporate Social Responsibility (CSR) and sustainable practices on a global level. These awards highlight the positive impact these organizations have on society, the environment, and their stakeholders worldwide. The applicability of these awards varies, and each award may have specific eligibility criteria. Here are some notable international CSR awards:

The CSR Excellence Awards: Organized by the Corporate Social Responsibility Institute, these awards recognize outstanding CSR initiatives globally. The applicability is generally open to companies, NGOs, and individuals from around the world.

Ethical Corporation Awards: These awards honor companies for their sustainability, CSR, and responsible business practices. They typically attract entries from various industries and regions globally.

Global Good Awards: Formerly known as the National CSR Awards UK, this event recognizes businesses and organizations worldwide for their positive impact on society and the environment. Applicability extends to international companies and organizations.

Corporate Knights Global 100: This award identifies the top 100 most sustainable companies globally based on several environmental and social indicators. It is applicable to companies from around the world.

The Oslo Business for Peace Award: Presented in Norway, this award honors business leaders who have demonstrated ethical and responsible behavior in their business practices. The award is open to international business leaders.

 The Comprehensive Guide On CSR For Consultant

The United Nations Global Compact Awards: These awards recognize businesses for their commitment to the UN Global Compact's ten principles on human rights, labor, environment, and anti-corruption. The applicability extends to companies participating in the UN Global Compact initiative.

FT/IFC Transformational Business Awards: Organized by the Financial Times (FT) and the International Finance Corporation (IFC), these awards celebrate innovative private-sector-led solutions to global development challenges. They are open to companies and organizations from around the world.

The Responsible Business Awards (formerly known as the Ethical Corporation Awards): These awards celebrate companies and leaders for their sustainable and responsible business practices. The applicability is generally open to international organizations.

The Global Sourcing Association (GSA) Awards: While primarily focused on the sourcing industry, these awards also recognize CSR and sustainability initiatives within the sector. The applicability extends to international sourcing companies and service providers.

Edie Sustainability Leaders Awards: These awards celebrate businesses and organizations that are leading the way in sustainability. The applicability is generally open to companies from around the world.

Ethical Corporation Responsible Business Awards: These awards focus on celebrating businesses that have integrated responsible and sustainable practices into their

   The Comprehensive Guide On CSR For Consultant

core strategies. Applicability is open to companies from various industries worldwide.

The Prince's Responsible Business Network Awards: Organized by Business in the Community (BITC) in the UK, these awards recognize businesses that have demonstrated responsible and sustainable practices. They are open to international companies and organizations that have a presence in the UK.

Corporate Responsibility Magazine's 100 Best Corporate Citizens: This annual list ranks companies based on their performance in various CSR and sustainability categories. Applicability is open to companies worldwide.

Sustainable Brands Awards: These awards celebrate companies that are leading the way in sustainability and responsible business practices globally. Applicability is open to international companies from various industries.

Responsible CEO of the Year Awards: Presented by Ethical Corporation, these awards honor CEOs who have demonstrated outstanding leadership in promoting CSR and sustainability. The applicability is generally open to CEOs of international companies.

The International CSR Excellence Awards: These awards recognize CSR initiatives that have made a significant positive impact on society and the environment globally. Applicability is open to companies, NGOs, and individuals from around the world.

Global Corporate Sustainability Awards (GCSA): These awards celebrate companies for their efforts in integrating

 The Comprehensive Guide On CSR For Consultant

sustainability into their business practices. Applicability is open to companies from various industries worldwide.

Global Good Governance Awards (3G Awards): While primarily focused on governance, these awards also consider CSR initiatives as an essential aspect of responsible business practices. Applicability is open to companies and organizations from around the world.

The European CSR Awards: These awards recognize outstanding CSR initiatives implemented by businesses, NGOs, and public sector organizations across Europe. Applicability is open to organizations operating in European countries.

The Global CSR Summit & Awards: These awards celebrate companies and leaders for their exceptional CSR efforts and positive social impact worldwide.

The Global Corporate Social Responsibility Impact Awards: These awards honor organizations for their exceptional CSR programs that have a positive impact on society and the environment worldwide. Applicability is open to companies, NGOs, and social enterprises globally.

The Ethical Corporation Awards for Excellence: These awards recognize companies for their efforts in embedding ethical practices, sustainability, and responsible business strategies. Applicability is generally open to international organizations from various sectors.

The World's Most Ethical Companies: This is an annual recognition given by the Ethisphere Institute to companies that demonstrate ethical leadership and responsible business practices. Applicability is open to companies worldwide.

The Global Good Fund's Global Good Awards: These awards celebrate social entrepreneurs and organizations driving positive social change. Applicability is open to social enterprises and individuals globally.

The Global Sustain Sustainability Awards: These awards recognize companies and organizations for their sustainability and CSR efforts, focusing on environmental, social, and governance initiatives. Applicability is open to international companies from various industries.

The Asia Responsible Enterprise Awards (AREA): While primarily focused on the Asia-Pacific region, these awards also consider companies' CSR and sustainable practices globally, provided they have a presence or significant impact in the Asia-Pacific region.

The Responsible Business Awards by Business in the Community (BITC) Ireland: These awards celebrate businesses for their commitment to responsible practices and sustainability in Ireland and beyond.

The Responsible Business Awards by Business in the Community (BITC) Cymru (Wales): These awards recognize responsible business practices and sustainability initiatives by companies operating in Wales and, in some categories, across the UK.

The Global Responsible Business Leadership Awards: These awards celebrate leaders who have demonstrated exemplary commitment to CSR and sustainability practices on a global scale.

 The Comprehensive Guide On CSR For Consultant

The Green World Awards: While primarily focused on environmental initiatives, these awards also consider CSR efforts that promote sustainability and ecological responsibility. Applicability is open to companies and organizations worldwide.

Please note that the applicability and eligibility criteria for these awards may change over time. Organizations interested in applying for any of these international CSR awards should refer to the respective award organizers' websites for the latest details, application guidelines, and deadlines. Additionally, new awards may have been introduced since my last update, so it's always beneficial to research and explore the latest options for recognition of CSR efforts on an international level.

The Comprehensive Guide On CSR For Consultant

Annexure 21: Employee Engagement In Corporate Social Responsibility Initiatives Example

CSR initiatives are most effective when employees actively participate and contribute to the cause. Here's a step-by-step example of how employees can be engaged in CSR initiatives:

CSR Initiative: "Education for All" - Providing Stationery Kits to Underprivileged Children

Needs Assessment and Objective Setting:

Identify the need for educational support among underprivileged children in a local community.

Set the objective of providing stationery kits to 100 children for the upcoming academic year.

Employee Survey and Involvement:

Conduct an internal survey to gauge employees' interest in supporting educational initiatives.

Form a CSR Committee comprising employees from different departments who are passionate about education and community development.

Collaboration with Local NGOs:

Connect with local NGOs or schools working with underprivileged children to understand their requirements and ensure a targeted approach.

Fundraising Campaign:

Organize a fundraising campaign within the organization to gather financial support for purchasing stationery kits.

Encourage employees to contribute voluntarily to the campaign.

Creating Awareness and Advocacy:

Hold awareness sessions or workshops on the importance of education for underprivileged children.

Invite guest speakers from NGOs or beneficiaries to share their experiences.

Stationery Kit Selection:

Involve employees in selecting the items to be included in the stationery kit.

Consider eco-friendly and sustainable options, aligning with the organization's values.

Assembling the Stationery Kits:

Organize a team-building activity where employees come together to assemble the stationery kits.

Foster a sense of camaraderie and shared purpose during this activity.

Distribution Event:

Plan a distribution event at the partner school or NGO facility.

Invite employees to join the event and interact with the children and teachers.

Employee Volunteering:

Offer employees the opportunity to volunteer their time to engage with the children, conduct workshops, or offer mentoring sessions.

Feedback and Impact Assessment:

Gather feedback from employees about their experiences and suggestions for improvement.

Conduct an impact assessment to measure the positive outcomes of the initiative, such as increased school attendance and improved academic performance.

Recognition and Appreciation:

Acknowledge and appreciate employees' contributions during company-wide meetings or through internal communication channels.

Highlight the collective impact achieved through the CSR initiative.

Continued Engagement:

Encourage ongoing engagement by sharing success stories and updates on the children's progress throughout the academic year.

Consider extending the initiative beyond one-time support to establish a long-term relationship with the partner organization.

By engaging employees in every step of the CSR initiative, the organization fosters a sense of ownership and shared responsibility for creating positive social impact. Employee involvement not only benefits the targeted beneficiaries but

 The Comprehensive Guide On CSR For Consultant

also enhances employee morale, teamwork, and company culture.

Employee engagement in Corporate Social Responsibility (CSR) initiatives is crucial for creating a positive impact both within the organization and in the communities, they serve. Here's an example of how employees can actively participate in CSR initiatives:

Example : Volunteer Day for Environmental Cleanup

Preparation and Communication:

The company communicates to all employees about an upcoming CSR initiative, a Volunteer Day for Environmental Cleanup, which aligns with the company's commitment to sustainability.

The HR department sends out a company-wide email, inviting employees to participate voluntarily in the event.

Task Assignment and Training:

A CSR team, led by HR and CSR representatives, organizes the event logistics and task assignments.

Employees are encouraged to sign up for specific tasks, such as collecting litter, planting trees, or conducting an awareness campaign.

Training and Safety Measures:

Employees involved in specific tasks receive training on safety protocols and environmental best practices.

They are provided with necessary safety gear, such as gloves and garbage bags.

Volunteer Day:

On the designated Volunteer Day, employees gather at a local park or beach, chosen for the cleanup.

The CSR team briefs the volunteers about the importance of their work and the impact of their efforts on the environment.

Environmental Cleanup Activities:

Employees split into groups and actively engage in various cleanup activities, collecting trash, segregating recyclables, and ensuring proper disposal.

Some employees engage in tree planting activities to enhance the green cover.

Awareness Campaign:

A group of volunteers sets up an awareness campaign station to educate passersby about environmental issues, the importance of waste management, and steps they can take to contribute to sustainability.

Documentation and Social Media:

The CSR team documents the entire event through photographs and videos.

The company's social media channels showcase the event, acknowledging the employees' efforts and creating awareness about the importance of CSR.

Post-Event Reflection:

    The Comprehensive Guide On CSR For Consultant

After the event, the company organizes a reflection session, where employees share their experiences and the impact they believe they made.

This session helps reinforce the importance of CSR and fosters a sense of pride and ownership among the employees.

Recognition and Rewards:

The company recognizes the employees' active participation in the CSR initiative, acknowledging their efforts during company-wide meetings or through internal newsletters.

Exceptional contributors may be given special recognition or rewards to incentivize ongoing engagement.

Employee engagement in CSR initiatives not only benefits the communities and the environment but also strengthens employee morale, teamwork, and company culture. It demonstrates the company's commitment to making a positive difference, both internally and externally.

   The Comprehensive Guide On CSR For Consultant

Annexure 22: Participatory Rural Appraisal (PRA) Techniques

The roots of PRA can be traced back to various sources, including the work of pioneers such as Robert Chambers, Jules N. Pretty, and Ian Scoones. These individuals were influenced by the ideas of Participatory Development and Rapid Rural Appraisal (RRA), which emphasized the importance of involving local communities in the development process.

The term "Participatory Rural Appraisal" was first coined by Robert Chambers, a development practitioner and researcher, in the early 1990s. Chambers advocated for a shift in development approaches, arguing that communities should be active participants in identifying their problems, setting priorities, and implementing solutions.

Participatory Rural Appraisal (PRA) is a set of techniques and methods used to involve local communities in the assessment, planning, and decision-making processes of rural development projects. PRA techniques are participatory and inclusive, empowering community members to actively contribute their knowledge, perspectives, and experiences. Here are some common PRA techniques:

Village Mapping: Community members draw maps of their village, depicting landmarks, resources, and infrastructure. It helps in understanding the spatial layout and resources of the community.

 The Comprehensive Guide On CSR For Consultant

Social Mapping: This technique involves creating a map that identifies social groups within the community, their roles, and relationships. It helps to understand social structures and power dynamics.

Transect Walks: Community members and facilitators walk through the village or surrounding areas, observing and documenting different features such as land use, natural resources, and infrastructure. It provides insights into the local environment.

Seasonal Calendars: Community members create calendars to show the different activities and events that occur during different seasons. It helps in understanding the seasonal variations in livelihoods and resource use.

Resource Mapping: Community members map local natural resources like water sources, forests, farmland, and grazing areas. It helps in sustainable resource management and planning.

Livelihood Analysis: A comprehensive analysis of the community's livelihood strategies, including agricultural practices, livestock rearing, and other income-generating activities.

Focus Group Discussions: Small groups of community members discuss specific topics or issues, encouraging open dialogue and capturing diverse perspectives.

Ranking and Scoring: Community members rank or score various options or priorities concerning development issues, helping in decision-making and project planning.

Seasonal Analysis: Examining how different factors, such as weather patterns, affect the community's livelihoods and resources throughout the year.

Community Timeline: Creating a timeline to trace the community's historical events and changes, providing context for current issues and challenges.

Community Wealth Ranking: Prioritizing households based on their wealth and vulnerability to target interventions effectively.

Matrix Ranking: Community members use a matrix to rank items based on specific criteria, such as project preferences or resource allocation.

PRA techniques are flexible and can be adapted to suit the needs and cultural context of the community. They are valuable tools for participatory development, fostering ownership and sustainability in rural development projects.

 The Comprehensive Guide On CSR For Consultant

Annexure 23: Nils Theory of Change

The Theory of Change (ToC) is a dynamic and systematic approach to social influence, development, and program evaluation. It is a comprehensive roadmap or visual picture of how an organization or program feels change occurs and the techniques it takes to attain its goals. The ToC defines the long-term goals, the paths to those goals, and the underlying assumptions that guide the intervention. It promotes clarity, transparency, and effectiveness in the design, implementation, and assessment of social programs by assisting stakeholders in understanding the causal links between actions, outputs, results, and impact.

Challenge: To challenge existing social issues, start by understanding the root causes and affected communities deeply. Set clear, measurable goals for change, and build awareness through storytelling and data. Collaborate with stakeholders, including experts, NGOs, government agencies, and the affected community. Develop innovative and sustainable solutions, advocate for policy changes, and engage in grassroots activism. Foster a culture of inclusivity, empathy, and diversity to drive collective action and address the social issue from multiple angles. Continuous monitoring, evaluation, and adaptation of strategies will ensure a more impactful and lasting transformation in society.

Change: Changing existing social issues requires a multifaceted approach. Start by raising awareness about the issue and its impact on affected communities. Engage

 The Comprehensive Guide On CSR For Consultant

stakeholders, including government, NGOs, and the private sector, to collaborate on solutions. Advocate for policy changes and implement targeted interventions to address root causes. Empower and involve affected communities in decision-making and program implementation. Foster a culture of inclusivity, equity, and empathy to drive sustainable change. Continuous monitoring and evaluation of efforts will help adapt strategies and ensure a lasting positive impact on society.

Sustain: Sustaining change in existing social issues requires long-term commitment and strategic efforts. Build strong partnerships with stakeholders to ensure collective ownership and support for ongoing initiatives. Foster community engagement and participation to empower local actors in driving change. Establish effective monitoring and evaluation systems to track progress and identify areas for improvement. Continuously adapt strategies based on lessons learned and changing circumstances. Invest in capacity building and knowledge transfer to empower individuals and institutions to continue the work. Advocate for supportive policies and allocate resources for sustainability. Cultivate a culture of social responsibility and foster collaboration between diverse sectors to create a lasting positive impact on society.

Example of a Social Issue in Challenge, Change, and Sustain Structure:

Corporate Social Responsibility Project Proposal: Promoting Digital Literacy in Rural Communities

Challenge: The challenge is to address the digital divide in rural communities, where limited access to technology and digital skills hinder socio-economic progress. Lack of digital literacy limits opportunities for education, employment, and access to essential services, perpetuating inequality and limiting community development.

Change: The project aims to promote digital literacy and bridge the digital divide in rural areas. It proposes establishing community-based digital centers equipped with computers, internet connectivity, and educational resources. Trained facilitators will provide digital skills training, including basic computer operations, internet usage, and online safety, empowering community members to leverage technology for personal and economic development.

Sustain: To sustain the project's impact, a multi-pronged approach will be adopted. Local partnerships with educational institutions, government agencies, and NGOs will facilitate resource sharing and continued training. Empowering local volunteers to serve as digital mentors will promote peer learning and support. Sustainability will be ensured by integrating digital literacy training into school curriculums and adult education programs. Regular monitoring and evaluation will measure the project's success and inform adaptations for greater effectiveness. Fundraising efforts and corporate partnerships will secure resources for long-term sustainability, allowing the project to continue empowering rural communities through digital literacy, ultimately leading to positive social and economic change.

Corporate Social Responsibility Project Proposal: Empowering Youth Through Skill Development for Employment

Challenge: The challenge is to address the high youth unemployment rate in a specific region, where many young individuals lack the necessary skills and training to access decent job opportunities. This issue leads to social and economic challenges, including increased poverty and disengagement from the workforce.

Change: The project proposes establishing a skill development and training center in the community. The center will offer various vocational courses aligned with local industry needs, such as computer programming, carpentry, culinary arts, and customer service. The training will be complemented with soft skills development, including communication, teamwork, and problem-solving, to enhance employability.

Sustain: To sustain the project's impact, a holistic approach will be adopted. Collaborations with local businesses will provide job placement opportunities for trained youth. Continuous feedback and communication with employers will inform course improvements to meet industry demands effectively. Engaging local mentors and professionals as guest speakers will inspire and motivate the youth. The center will also seek funding from corporate sponsors, government grants, and philanthropic organizations to ensure long-term sustainability. Monitoring and evaluation mechanisms will track graduates' job placements and income levels to demonstrate the project's success and identify areas for improvement, thus creating a sustainable cycle of skill development and youth empowerment for lasting social change.

Social Issue: Education and Empowerment of Backward Classes

 The Comprehensive Guide On CSR For Consultant

Challenge: Chatrapati Shahu Maharaj faced the challenge of addressing the educational and social disparities faced by the backward classes in the princely state of Kolhapur, India. The caste system and social prejudices prevented backward class communities from accessing education and participating in decision-making. The challenge was to challenge the prevailing social norms and create a more inclusive and equitable society.

Change: Chatrapati Shahu Maharaj initiated transformative changes by implementing significant social and educational reforms during his reign from 1894 to 1922. He established more than 200 schools, including the Rajaram College, to provide education to backward class students. He encouraged and supported the education of women and made efforts to eradicate untouchability, fostering a more inclusive social environment.

Sustain: To sustain Chatrapati Shahu Maharaj's work, ongoing efforts are required to provide quality education and equal opportunities to the backward class communities. This involves ensuring access to education and skill development programs, promoting social justice and inclusivity, and continuing to challenge discriminatory practices. Sustaining his legacy requires collaborative efforts with government bodies, NGOs, and community organizations to advocate for policies and programs that empower the backward classes. Regular monitoring and evaluation of educational initiatives are essential to adapt strategies and drive lasting change, ensuring the sustained progress of education and empowerment for backward class communities.

Social Issue: Dalit Empowerment and Social Justice

The Comprehensive Guide On CSR For Consultant

Challenge: Dr. Babasaheb Ambedkar faced the significant challenge of addressing the deep-rooted social discrimination and oppression faced by Dalits in India. The caste-based hierarchy and prevailing discriminatory practices hindered their access to education, economic opportunities, and basic human rights. The challenge was to challenge the existing social norms and create a more equitable society.

Change: Dr. Ambedkar initiated several transformative changes to uplift Dalits. He advocated for their rights and worked tirelessly to abolish untouchability and caste-based discrimination. He drafted the Indian Constitution, which enshrined principles of equality, social justice, and affirmative action to provide opportunities for the upliftment of marginalized communities, including Dalits.

Sustain: To sustain Dr. Ambedkar's work, ongoing efforts are required to enforce laws against discrimination and promote social inclusion. This involves continuing to advocate for policies that uplift Dalits, provide equal access to education and employment opportunities, and eradicate social prejudices. Sustaining Dr. Ambedkar's legacy requires fostering a society that respects and values diversity, challenges caste-based discrimination, and ensures the social, economic, and political empowerment of Dalits. Continuous education, awareness campaigns, and collaboration with civil society organizations are essential to drive lasting change and build a more inclusive and just society.

Social Issue: Women's Education and Empowerment

   The Comprehensive Guide On CSR For Consultant

Challenge: Mahashri Karve faced the challenge of addressing the prevalent gender disparities and limited access to education for women in India. The social norms and patriarchal traditions posed significant obstacles to women's education and empowerment. The challenge was to challenge these norms and create opportunities for women to pursue education and lead independent lives.

Change: Mahashri Karve initiated transformative changes by establishing the first women's university in India, the SNDT Women's University, in 1916. This pioneering step provided women with access to higher education and opened doors for their empowerment. She also founded numerous educational institutions, vocational training centers, and hostels to support women's education and skill development.

Sustain: To sustain Mahashri Karve's work, ongoing efforts are required to promote women's education and empowerment. This involves continuing to provide access to quality education and vocational training for women, especially in rural and marginalized communities. Sustaining her legacy requires challenging traditional gender roles, advocating for gender-sensitive policies, and empowering women to participate in decision-making at all levels. Collaboration with government agencies, NGOs, and community-based organizations is crucial to driving lasting change and ensuring the sustained progress of women's education and empowerment. Continuous monitoring and evaluation of programs will help adapt strategies and maximize the impact of her work.

Social Issue: Gender Inequality

Challenge: The challenge is to address the systemic gender inequality that exists in various aspects of society, such as education, employment, and representation in decision-making positions. This involves recognizing and understanding the deeply rooted social norms and biases that perpetuate gender disparities.

Change: The change phase focuses on implementing targeted interventions and initiatives to promote gender equality. This may include campaigns to raise awareness about gender issues, advocating for equal opportunities in education and employment, promoting women's leadership and empowerment, and addressing harmful stereotypes and cultural practices.

Sustain: To sustain the change, efforts must be continuous and long-term. This involves building a supportive and inclusive environment that actively promotes gender equality and challenges discriminatory practices. Ensuring equal access to resources, opportunities, and services is crucial. Sustaining change also requires ongoing advocacy for policy reforms, institutional changes, and partnerships with diverse stakeholders to create a society where gender equality is firmly embedded in all aspects of life. Monitoring and evaluation help in measuring progress and adapting strategies to achieve lasting impact.

Social Issue: Environmental Pollution

Challenge: The challenge is to address the escalating environmental pollution caused by various human activities, such as industrial emissions, waste generation, deforestation, and excessive use of non-renewable resources. Understanding the scale of environmental degradation and its impacts on ecosystems, biodiversity, and human health is crucial to tackling this issue.

Change: The change phase focuses on implementing sustainable practices and policies to reduce pollution levels and mitigate environmental damage. This includes promoting renewable energy sources, encouraging waste reduction and recycling, advocating for stricter regulations on industrial emissions, and fostering eco-friendly practices in agriculture and transportation.

Sustain: To sustain the change, continuous efforts are required to instill a culture of environmental responsibility and consciousness. This involves raising awareness about individual and collective actions that contribute to pollution and fostering a sense of stewardship towards the environment. Building strong partnerships between governments, industries, communities, and environmental organizations is essential for the implementation of long-term solutions. Regular monitoring of pollution levels and environmental health is necessary to ensure progress and adapt strategies accordingly. Creating incentives for sustainable practices and enforcing environmental regulations play a vital role in sustaining positive change and preserving the environment for future generations.

Social Issue: Environmental Sustainability

Challenge: The challenge for a company engaging in corporate social responsibility (CSR) is to address its environmental impact and contribute to environmental sustainability. This may involve reducing carbon emissions, minimizing waste generation, conserving natural resources,

 The Comprehensive Guide On CSR For Consultant

and mitigating the negative effects of its operations on the environment.

Change: The change phase involves implementing various sustainable practices and initiatives. The company can invest in renewable energy sources, adopt eco-friendly production processes, promote recycling and waste reduction, and implement green supply chain practices. By integrating sustainability into its business model, the company aims to minimize its ecological footprint and contribute positively to the environment.

Sustain: To sustain the CSR efforts related to environmental sustainability, the company must embed sustainability in its corporate culture and long-term strategies. This involves setting ambitious sustainability goals, engaging stakeholders (employees, suppliers, customers, and local communities) in the process, and promoting environmental consciousness among employees and customers. Continuous monitoring and reporting on sustainability performance are essential to track progress and ensure the company's commitment to environmental sustainability is maintained over time. Additionally, collaborating with external stakeholders and participating in collective sustainability initiatives can enhance the company's impact and contribute to the broader sustainability movement.

 The Comprehensive Guide On CSR For Consultant

Annexure 24: Small and Medium Enterprises (SMEs) can develop the CSR and Sustainability Practices

Small and Medium Enterprises (SMEs) can collectively perform the best CSR and sustainability practices by collaborating and sharing resources, knowledge, and experiences. Here are some ways in which SMEs can work together:

Collaborative Initiatives:

Establish local networks: SMEs can form local networks or associations focused on CSR and sustainability. This allows for sharing best practices, knowledge, and resources among members.

Joint projects: SMEs can collaborate on CSR projects, such as environmental conservation, community development, or social impact initiatives. Pooling resources and expertise can maximize the impact of their efforts.

Supplier partnerships: SMEs can prioritize working with suppliers that align with their CSR and sustainability goals, encouraging responsible practices throughout the supply chain.

Shared Resources:

Training and capacity-building: SMEs can organize joint training programs or workshops to enhance their understanding of CSR, sustainability practices, and implementation strategies.

 The Comprehensive Guide On CSR For Consultant

Shared expertise: SMEs can share their experiences, challenges, and successes in implementing CSR and sustainability initiatives. This can be done through knowledge-sharing sessions, webinars, or industry forums.

Sharing tools and resources: SMEs can collaborate to develop shared resources such as CSR guidelines, templates, or toolkits that provide practical guidance for implementing sustainability practices.

Advocacy and Influence:

Collective voice: SMEs can join forces to advocate for policies and regulations that promote sustainability, CSR incentives, or support for small businesses engaging in responsible practices.

Industry standards and certifications: SMEs can work together to establish industry-specific CSR and sustainability standards and certifications that align with their sector's unique needs and challenges.

Reporting and Transparency:

Sharing impact reports: SMEs can collaborate to develop joint CSR and sustainability impact reports, showcasing their collective efforts and demonstrating their commitment to responsible business practices.

Transparency in supply chains: SMEs can collectively work towards improving transparency in supply chains, sharing information on sourcing practices, responsible production, and ethical procurement.

Engaging with Local Communities:

 The Comprehensive Guide On CSR For Consultant

Community partnerships: SMEs can collaborate with local communities on social development projects, such as education, healthcare, or skills training. This promotes shared value and addresses community needs.

Volunteering and employee engagement: SMEs can encourage their employees to participate in community service activities collectively, such as volunteering for local charities or supporting environmental cleanup campaigns.

Sustainable Innovation:

Collaborative research and development: SMEs can collaborate on research and development projects focused on sustainable technologies, products, or services that benefit the environment and society.

Knowledge sharing on innovation: SMEs can share information and experiences on sustainable innovation, promoting the adoption of environmentally friendly and socially responsible practices within their industries.

By working together, SMEs can amplify their impact, share resources and expertise, and create a collective force for positive change. Collaboration among SMEs enables them to overcome individual limitations and contribute significantly to CSR and sustainability practices.

 The Comprehensive Guide On CSR For Consultant

Annexure 25: Best CSR Practices of listed companies

Please keep in mind that CSR programmes might change over time, so check the official website or recent reports for the most up-to-date information on their CSR efforts.

1. Reliance Industries Limited (RIL),

RIL an Indian conglomerate, was involved in a number of important Corporate Social Responsibility (CSR) initiatives. Please keep in mind, however, that these initiatives may have evolved or modified since then. Here are some of their notable CSR actions up to that point:

Reliance Foundation: The philanthropic arm of Reliance Industries, Reliance Foundation, has been actively involved in various social and community development programs. Some key focus areas include healthcare, education, rural development, and disaster response.

Education and Skill Development: Reliance Foundation has initiated several programs to improve access

to quality education and skill development opportunities. This includes establishing schools, vocational training centers, and scholarships for deserving students.

Rural Transformation: RIL's CSR initiatives have targeted rural development and upliftment. The company has undertaken projects to enhance agricultural practices, support farmers, and create sustainable livelihood opportunities in rural areas.

 The Comprehensive Guide On CSR For Consultant

Healthcare Initiatives: RIL has been actively supporting healthcare initiatives, including building and upgrading hospitals, providing medical equipment, and organizing health camps for underserved communities.

Disaster Relief: Reliance Foundation has been quick to respond to natural disasters and emergencies by providing aid and relief materials to affected communities.

Swachh Bharat Abhiyan: RIL actively participated in the Swachh Bharat (Clean India) campaign initiated by the Government of India. They have undertaken cleanliness drives and promoted sanitation awareness.

Environmental Sustainability: The company has been working towards reducing its environmental impact by implementing sustainable practices, promoting renewable energy, and contributing to environmental conservation efforts.

Digital Education: RIL has been involved in initiatives that promote digital education and digital literacy, aiming to bridge the digital divide and empower individuals with digital skills.

Community Development: RIL has been actively engaged in community development projects, focusing on infrastructure development, water conservation, and social empowerment.

2. Tata Consultancy Services Limited (TCS)

TCS is well-known for its significant commitment to Corporate Social Responsibility (CSR) projects. TCS has been active in a variety of social and community development programmes, with a focus on education,

 The Comprehensive Guide On CSR For Consultant

healthcare, environmental sustainability, and skill development. While specific programmes may have evolved or new projects launched after then, the following are some of their notable CSR efforts up to that point:

TCS' Ignite: TCS Ignite is a program focused on enhancing the quality of education in schools, especially in rural and underprivileged areas. The program provides digital and interactive learning tools, teacher training, and infrastructure support to improve the overall learning experience for students.

Adult Literacy: TCS has initiated adult literacy programs to provide basic education and functional literacy skills to adults from disadvantaged communities. This helps enhance their employability and overall quality of life.

TCS IT Wiz: The TCS IT Wiz is an annual technology quiz competition designed to encourage students to develop an interest in information technology and computer science. The competition is conducted across various cities and schools in India, promoting digital awareness and tech literacy.

TCS Health and Nutrition Program: This initiative focuses on improving the health and well-being of underserved communities. TCS partners with NGOs and local authorities to provide healthcare facilities, nutrition support, and awareness programs on health-related issues.

Go Green Initiatives: TCS has undertaken various environmental sustainability initiatives, such as waste management, energy conservation, and promoting renewable energy sources. They also engage in tree

 The Comprehensive Guide On CSR For Consultant

plantation drives to contribute to environmental conservation.

Skill Development: TCS has been actively involved in skill development programs to empower youth with the necessary skills for employment. They offer vocational training and digital literacy programs to enhance employability.

Rural Development: TCS undertakes projects to support rural development and upliftment. This includes infrastructure development, access to clean water, and rural entrepreneurship programs.

Employee Volunteering: TCS encourages its employees to actively participate in CSR initiatives through its 'Employee Volunteerism' program. This program allows employees to contribute their time and skills to various social causes and community development projects.

3. HDFC Bank Limited,

HDFC Bank Limited is one of India's leading private sector banks, was involved in various Corporate Social Responsibility (CSR) initiatives. While I cannot provide real-time information on their most recent initiatives, I can highlight some of their noteworthy CSR efforts up until that time:

HDFC Educational Crisis Scholarship Support (ECSS): The bank provides scholarships to students facing financial crises and struggling to continue their education. This initiative helps deserving students access quality education and create a better future for themselves.

   The Comprehensive Guide On CSR For Consultant

Parivartan: HDFC Bank's Parivartan program focuses on rural development and sustainable livelihoods. Through various projects, the bank empowers rural communities by promoting education, health, skill development, and rural infrastructure.

Swachhagraha: This initiative is aimed at promoting cleanliness and sanitation across the country. The bank collaborates with local authorities and communities to organize cleanliness drives, awareness campaigns, and waste management programs.

Sustainable Livelihood Initiative: Under this program, HDFC Bank supports underprivileged communities and women by providing skill development training and livelihood opportunities, enabling them to become self-reliant.

HDFC Bank SmartBuy Donation Platform: The bank's online platform, HDFC Bank SmartBuy, allows customers to make donations to various NGOs and charitable organizations, thereby facilitating individual contributions to social causes.

Blood Donation Drives: HDFC Bank organizes regular blood donation camps across its branches and offices to contribute to the health and well-being of society.

Support to Healthcare: The bank extends support to hospitals, medical centers, and NGOs working in the healthcare sector to improve medical facilities and reach underserved communities.

Environmental Initiatives: HDFC Bank has taken steps to reduce its carbon footprint and promote environmental

 The Comprehensive Guide On CSR For Consultant

sustainability. They have introduced green banking practices, energy-efficient measures, and tree plantation drives.

4. Infosys Limited

Infosys Limited, one of India's leading IT services companies, has been actively involved in various Corporate Social Responsibility (CSR) initiatives. While specific initiatives may have evolved or new projects may have been launched since then, here are some of their noteworthy CSR efforts up until that time:

Infosys Foundation: The Infosys Foundation, the philanthropic arm of Infosys, has been at the forefront of various social initiatives. Some key focus areas of the foundation include education, healthcare, rural development, arts and culture, and destitute care.

Education and Digital Literacy: Infosys Foundation has undertaken several initiatives to promote education and digital literacy. They have set up schools, computer centers, and libraries, especially in rural and underprivileged areas, to improve access to quality education and technology.

Healthcare Initiatives: Infosys Foundation has been actively involved in supporting healthcare initiatives. They have contributed to the development of medical facilities, provided medical equipment to hospitals, and organized health camps for underserved communities.

Rural Development: The foundation has undertaken projects to support rural development and upliftment. This includes infrastructure development, water conservation, and rural empowerment programs.

 The Comprehensive Guide On CSR For Consultant

Mid-Day Meal Program: Infosys Foundation has been supporting the Mid-Day Meal program, which aims to provide nutritious meals to schoolchildren to improve attendance and overall health.

Environmental Sustainability: Infosys has been working towards reducing its environmental impact by implementing sustainable practices and promoting green initiatives. The company has focused on energy efficiency, waste management, and renewable energy sources.

Disaster Relief: Infosys has responded promptly to natural disasters and emergencies by providing aid and relief materials to affected communities.

Employee Volunteering: Infosys encourages its employees to actively participate in CSR initiatives through its 'Infosys Volunteers' program. This program allows employees to contribute their time and skills to various social causes and community development projects.

Empowering Persons with Disabilities (PWD): Infosys Foundation has been involved in initiatives to empower persons with disabilities by providing them with vocational training, employment opportunities, and accessibility support.

5. ICICI Bank Limited

ICICI Bank Limited, one of India's leading private sector banks, has been actively involved in various Corporate Social Responsibility (CSR) initiatives. While specific initiatives may have evolved or new projects may have been launched since then, here are some of their noteworthy CSR efforts up until that time:

ICICI Foundation for Inclusive Growth: The ICICI Foundation focuses on various social and economic development initiatives, including skill development and livelihood training programs. These programs aim to enhance the employability of youth from marginalized communities.

ICICI Academy for Skills: The bank has set up multiple training centers called "ICICI Academy for Skills" across India. These centers provide free vocational training to underprivileged youth in various trades, helping them secure jobs and become financially independent.

Financial Inclusion: ICICI Bank has been actively working towards promoting financial inclusion by reaching out to underserved and unbanked populations. The bank aims to bring them into the formal banking system through various outreach programs.

Education Support: ICICI Bank supports educational initiatives by providing scholarships and financial assistance to deserving students from economically weaker sections of society.

Health and Sanitation: The bank has undertaken various health and sanitation initiatives, including organizing health camps and supporting healthcare facilities for underprivileged communities.

Rural Development: ICICI Bank's CSR initiatives also focus on rural development. The bank undertakes projects related to agriculture, rural infrastructure, and community development in rural areas.

 The Comprehensive Guide On CSR For Consultant

Environmental Sustainability: ICICI Bank has been working towards reducing its environmental impact and promoting sustainable practices. They have implemented energy-efficient measures and green banking initiatives.

Women Empowerment: The bank supports programs that empower women economically and socially. This includes providing training and financial assistance to women entrepreneurs and supporting women-centric NGOs.

Disaster Relief: ICICI Bank actively contributes to relief efforts during natural disasters and emergencies by providing financial aid and support to affected communities.

6. Hindustan Unilever Limited (HUL),

Hindustan Unilever Limited (HUL), one of India's leading consumer goods companies, has been actively involved in various Corporate Social Responsibility (CSR) initiatives. While specific initiatives may have evolved or new projects may have been launched since then, here are some of their noteworthy CSR efforts up until that time:

Project Shakti: HUL's Project Shakti aims to empower rural women by providing them with entrepreneurial opportunities. The initiative trains and supports rural women as Shakti Ammas (entrepreneurs) who sell HUL products in their communities, creating livelihood opportunities for them.

Swachh Aadat Swachh Bharat: This initiative is part of HUL's commitment to the Swachh Bharat Abhiyan (Clean India Mission). It focuses on promoting cleanliness and

 The Comprehensive Guide On CSR For Consultant

hygiene by encouraging people to adopt good habits for personal and environmental health.

Water Conservation and Management: HUL is actively engaged in water conservation and management initiatives. The company has undertaken projects to replenish water resources, improve water access in water-stressed areas, and promote responsible water usage in their operations.

Project Prabhat: HUL's Project Prabhat focuses on enhancing the quality of education in rural schools. The initiative provides educational infrastructure, teacher training, and digital learning tools to improve the overall learning experience for students.

Project Sunlight: Project Sunlight is HUL's initiative aimed at inspiring sustainable living and positive social change. It promotes sustainable practices, environmental conservation, and community development.

HUL Foundation: The HUL Foundation, the philanthropic arm of the company, supports various social causes, including education, healthcare, and environmental sustainability.

Lifebuoy's Handwashing Campaigns: HUL's Lifebuoy brand has been at the forefront of promoting handwashing habits for improved public health. They have launched various awareness campaigns, especially during times of disease outbreaks.

Support to Rural Farmers: HUL has undertaken projects to support and empower rural farmers by providing training, access to technology, and fair market linkages for their agricultural produce.

 The Comprehensive Guide On CSR For Consultant

Sustainable Sourcing: HUL is committed to sustainable sourcing practices, which include promoting responsible agricultural and raw material sourcing to minimize the environmental impact of their products.

7. State Bank of India (SBI)

Bank of India (SBI), India's largest public sector bank, has been actively involved in various Corporate Social Responsibility (CSR) practices. While specific practices may have evolved or new initiatives may have been launched since then, here are some of their noteworthy CSR efforts up until that time:

Financial Inclusion: SBI has been working towards promoting financial inclusion by extending banking services to underprivileged and unbanked populations. The bank aims to bring them into the formal banking system through various outreach programs.

Education and Skill Development: SBI's CSR initiatives include supporting education and skill development programs. They have provided scholarships, built schools, and offered vocational training to empower youth from marginalized communities.

Health and Sanitation: The bank has undertaken various health and sanitation initiatives, including organizing health camps, supporting healthcare facilities, and promoting awareness of health-related issues.

Rural Development: SBI's CSR initiatives focus on rural development, including infrastructure development, support to agricultural practices, and community development in rural areas.

 The Comprehensive Guide On CSR For Consultant

Disaster Relief: SBI has actively contributed to relief efforts during natural disasters and emergencies by providing financial aid and support to affected communities.

Swachh Bharat Abhiyan: SBI has been actively participating in the Swachh Bharat (Clean India) campaign initiated by the Government of India. The bank has undertaken cleanliness drives and promoted sanitation awareness.

Environmental Sustainability: SBI has been working towards reducing its environmental impact and promoting sustainable practices. They have implemented energy-efficient measures and green banking initiatives.

Empowerment of Women: SBI supports initiatives that empower women economically and socially. This includes providing training and financial assistance to women entrepreneurs and supporting women-centric NGOs.

Skill Loan Scheme: As part of their CSR efforts, SBI launched the Skill Loan Scheme to provide financial assistance to individuals seeking skill development training, thereby improving their employability.

8. Housing Development Finance Corporation Limited (HDFC Ltd)

Housing Development Finance Corporation Limited (HDFC Ltd) has been actively involved in various Corporate Social Responsibility (CSR) and sustainability initiatives. While specific initiatives may have evolved or new projects may have been launched since then, here are some of their noteworthy CSR and sustainability efforts up until that time:

HDFC Education and Healthcare Initiatives: HDFC Ltd has been supporting various education and healthcare programs to improve access to quality education and healthcare services for underprivileged communities. They have contributed to the development of schools, scholarships for deserving students, and medical facilities.

Affordable Housing Initiatives: As a housing finance company, HDFC has focused on promoting affordable housing initiatives. They have collaborated with various stakeholders to create affordable housing projects and support the Pradhan Mantri Awas Yojana (PMAY) scheme in India.

Environmental Sustainability: HDFC has been working towards reducing its environmental impact and promoting sustainable practices. They have implemented energy-efficient measures in their offices and embraced green building practices.

Disaster Relief: HDFC has actively contributed to relief efforts during natural disasters and emergencies by providing financial aid and support to affected communities.

HDFC Skill Development Initiative: The company has undertaken skill development programs to enhance the employability of youth from marginalized communities. They provide vocational training and support for livelihood generation.

Employee Volunteering: HDFC encourages its employees to actively participate in CSR initiatives through its 'Parivartan' program. This program allows employees to contribute their time and skills to various social causes and community development projects.

Community Development: HDFC has been actively engaged in community development projects, focusing on infrastructure development, water conservation, and social empowerment.

HDFC Swachhagraha: As part of its CSR efforts, HDFC has launched the "HDFC Swachhagraha" initiative, aiming to promote cleanliness and hygiene by encouraging behavioral changes among citizens.

CSR Partnerships: HDFC has partnered with various NGOs and social organizations to amplify the impact of their CSR initiatives and address critical social challenges effectively.

9. Bharti Airtel Limited,

Bharti Airtel Limited, one of India's leading telecommunications companies, has been actively involved in various Corporate Social Responsibility (CSR) and sustainability initiatives. While specific initiatives may have evolved or new projects may have been launched since then, here are some of their noteworthy CSR and sustainability efforts up until that time:

Education Initiatives: Bharti Airtel has been supporting various education programs to improve access to quality education for underprivileged children. They have undertaken initiatives to provide scholarships, digital education resources, and educational infrastructure.

 The Comprehensive Guide On CSR For Consultant

Healthcare Initiatives: The company has been actively involved in supporting healthcare initiatives. Bharti Airtel has partnered with healthcare organizations to provide medical facilities, healthcare access, and awareness programs for communities in need.

Digital Inclusion: Bharti Airtel's CSR initiatives focus on promoting digital inclusion, especially in rural and remote areas. They have launched initiatives to provide internet connectivity and digital literacy programs to bridge the digital divide.

Environmental Sustainability: Bharti Airtel has been working towards reducing its environmental impact and promoting sustainable practices. They have implemented energy-efficient measures and green initiatives to minimize their carbon footprint.

Disaster Relief: Bharti Airtel has actively contributed to relief efforts during natural disasters and emergencies by providing communication support, financial aid, and support to affected communities.

Women Empowerment: The company supports initiatives that empower women economically and socially. This includes providing training and financial assistance to women entrepreneurs and supporting women-centric NGOs.

Empowering Youth: Bharti Airtel has undertaken skill development programs to empower youth with the necessary skills for employment. They offer vocational training and digital literacy programs to enhance employability.

 The Comprehensive Guide On CSR For Consultant

Community Development: The company has been actively engaged in community development projects, focusing on infrastructure development, sanitation, and social empowerment.

Employee Engagement: Bharti Airtel encourages its employees to actively participate in CSR initiatives through its 'Employee Volunteering' program. This program allows employees to contribute their time and skills to various social causes and community development projects.

Annexure 26: Carbon Footprint Reduction

Carbon footprint is the total quantity of greenhouse gas emissions, mostly CO2, produced directly and indirectly by an individual, organisation, event, or product throughout its existence. It measures human-induced climate change.

The carbon footprint is usually quantified in metric tonnes of CO2e. This unit compares the global warming potential of methane (CH4) and nitrous oxide (N2O) to carbon dioxide.

Carbon footprints consider emissions from:

Direct Emissions (Scope 1): Emissions from sources owned or controlled by the entity, such as burning fossil fuels for heating or driving.

Indirect Emissions (Scope 2): Emissions from the entity's energy, heat, or steam purchases and use.

Other Indirect Emissions (Scope 3): Supply chains, employee commuting, company travel, waste disposal, and product use and disposal are examples of indirect emissions.

 The Comprehensive Guide On CSR For Consultant

Energy, fuel, trash, travel, and other statistics are needed to calculate the carbon footprint. Emission factors transform data into CO2e, which is multiplied by quantities to calculate the carbon footprint.

Climate change requires measuring and reducing carbon footprints. It helps individuals, organisations, and governments assess their environmental effect and discover emissions reduction potential through sustainable practises, energy efficiency, renewable energy adoption, waste management, and other mitigation techniques.

To show how the method for calculating a smaller carbon footprint works, let's look at a made-up example of a company that has taken steps to save energy:

average Emissions: Based on how much energy the company uses, it figures that its average emissions are 1,000 metric tonnes of CO2 equivalent (tCO2e) per year.

Reduced Emissions: After putting in place steps to save energy, the company figures out that its energy use has cut its emissions by 800 tCO2e per year.

Reducing your carbon footprint: Using the formula, here's how to figure out how much your carbon impact will shrink:

Baseline emissions - reduced emissions = 1,000 tCO2e - 800 tCO2e = 200 tCO2e.

So, in this case, the company was able to reduce its carbon footprint by 200 metric tonnes of CO2 equivalent per year by making changes to save energy.

 The Comprehensive Guide On CSR For Consultant

Please keep in mind that this is a simple example meant to show how things work. In fact, figuring out how to reduce a company's carbon footprint requires looking at more data and taking into account sources of emissions like energy, transportation, waste, and more. To correctly measure and keep track of carbon footprint reductions, it is best to use specialised carbon accounting methods and tools.

Annexure 27: Corporate Social Responsibility Consultant

Becoming a professional Corporate Social Responsibility (CSR) consultant requires a combination of knowledge, skills, and experience. Here are some steps to help you pursue a career in CSR consulting:

Develop Expertise: Gain a deep understanding of CSR concepts, sustainability principles, and social impact strategies. Stay updated on emerging trends, best practices, and relevant frameworks such as the Global Reporting Initiative (GRI) or the United Nations Sustainable Development Goals (SDGs).

Gain practical experience by working in the field of CSR, sustainability, or related areas. This can include internships, entry-level positions, or volunteering with organizations that focus on CSR initiatives.

Seek opportunities to work on CSR projects within your current organization or community to gain hands-on

   The Comprehensive Guide On CSR For Consultant

experience in designing, implementing, and evaluating CSR initiatives.

Acquire Relevant Education: Consider pursuing a degree or certification program in fields related to CSR, sustainability, business ethics, or social impact. This formal education will provide you with a strong foundation and specialized knowledge.

Obtain a relevant degree in fields such as business administration, sustainability, social sciences, or environmental studies. This foundation will provide you with a solid understanding of CSR principles, sustainability practices, and ethical considerations.

Stay updated on current CSR trends, best practices, and emerging issues through continuous learning, attending conferences, webinars, and participating in professional networks.

Build a Diverse Skill Set: Develop a range of skills that are essential for CSR consulting. These may include project management, data analysis, stakeholder engagement, strategic planning, communication, and corporate governance.

Hone your consulting skills, including problem-solving, project management, strategic thinking, communication, and facilitation. These skills are crucial for effectively working with clients and stakeholders.

Gain experience in conducting assessments, developing CSR strategies, designing and implementing programs, and measuring social impact.

 The Comprehensive Guide On CSR For Consultant

Gain Practical Experience: Seek opportunities to gain practical experience in the field of CSR. This can include internships, volunteering with nonprofit organizations, or working on CSR projects within companies. Build a portfolio of relevant projects and outcomes that demonstrate your expertise and impact.

Consider specializing in specific CSR areas, such as environmental sustainability, community development, or employee engagement, to develop expertise and differentiate yourself in the market.

Stay informed about the latest CSR frameworks, standards, and reporting guidelines, such as the Global Reporting Initiative (GRI) or ISO 26000, to provide comprehensive and up-to-date advice to clients.

Networking and Professional Development: Attend industry conferences, seminars, and workshops to expand your professional network and stay connected with CSR practitioners. Join professional associations and organizations related to CSR and sustainability to access resources, training, and networking opportunities.

Connect with professionals in the CSR field through industry events, networking platforms, and social media. Attend conferences, workshops, and seminars to meet experts and learn from their experiences.

Join professional associations and organizations focused on CSR, sustainability, and corporate citizenship to expand your network and access resources and knowledge.

Stay Informed: Continuously update your knowledge and skills by reading industry publications, research reports,

 The Comprehensive Guide On CSR For Consultant

and academic journals. Engage in online forums, webinars, and discussions to stay informed about the latest trends and challenges in CSR.

Develop Consultancy Skills: Enhance your consulting skills, including client management, problem-solving, facilitation, and presentation skills. Learn how to effectively communicate complex CSR concepts to diverse stakeholders and develop tailored solutions for clients.

Build a portfolio of successful CSR projects and initiatives that showcase your expertise and the positive outcomes you have achieved.

Use case studies, reports, and testimonials to demonstrate the value you bring as a CSR consultant and the impact you have made on organizations and communities.

Market Yourself: Create a professional brand by developing a strong online presence, including a website, blog, and social media profiles that showcase your expertise and thought leadership in CSR. Develop a network of contacts and clients through referrals, partnerships, and attending industry events.

Ethics and Integrity: Maintain the highest standards of ethics and integrity in your work as a CSR consultant. Demonstrate a commitment to social and environmental responsibility, and align your own practices with the principles you advocate.

Adhere to high ethical standards, ensuring transparency, integrity, and accountability in your consulting practice.

 The Comprehensive Guide On CSR For Consultant

Stay objective and provide unbiased advice that aligns with the best interests of your clients and the broader social and environmental context.

Continual Learning: Commit to lifelong learning and professional development. Stay updated on evolving CSR trends, emerging frameworks, and new tools and technologies that can enhance your consulting services.

By combining knowledge, skills, practical experience, and a passion for driving positive social impact, you can position yourself as a professional CSR consultant ready to help organizations navigate the complex world of corporate social responsibility.

 The Comprehensive Guide On CSR For Consultant

Annexure 28: References

1. https://www.tatatrusts.org/
2. https://www.gatesfoundation.org/
3. https://azimpremjifoundation.org/
4. https://www.reliancefoundation.org/
5. https://www.rockefellerfoundation.org/
6. https://www.carnegiefoundation.org/
7. https://www.thehowardgbuffettfoundation.org/
8. https://www.opensocietyfoundations.org/
9. https://www.thehersheycompany.com/en_us/home/about-us/milton-hershey.html
10. https://en.wikipedia.org/wiki/Marta_Abreu
11. https://www.oprahfoundation.org/
12. https://chanzuckerberg.com/
13. https://www.shivnadarfoundation.org/
14. https://ms-mf.org/
15. https://unglobalcompact.org/
16. https://www.iso.org/iso-26000-social-responsibility.html

17. https://www.undp.org/sustainable-development-goals/
18. https://www.globalreporting.org/standards/
19. https://www.oecd.org/corporate/mne/
20. https://www.ilo.org/global/standards/introduction-to-international-labour-standards/conventions-and-recommendations/lang--en/index.htm
21. https://www.cdp.net/en
22. https://www.ohchr.org/sites/default/files/documents/publications/guidingprinciplesbusinesshr_en.pdf
23. https://www.accountability.org/standards/aa1000-assurance-standard/
24. https://sa-intl.org/resources/sa8000-standard/
25. https://www.oecd.org/corporate/mne/WP-2001_5.pdf
26. http://www.socialvaluelab.org.uk/wp-content/uploads/2016/09/SROI-a-guide-to-social-return-on-investment.pdf
27. https://www.lbg-canada.ca/the-lbg-model/
28. https://www.mca.gov.in/Ministry/latestnews/National_Voluntary_Guidelines_2011_12jul2011.pdf
29. https://online.hbs.edu/blog/post/what-is-the-triple-bottom-line
30. https://www.bsr.org/en/reports/stakeholder-engagement-five-step-approach-toolkit
31. https://www.unpri.org/
32. https://www.wbcsd.org/Overview/About-us/Vision-2050-Time-to-Transform
33. https://thenaturalstep.org/
34. https://www.unilever.com/
35. https://www.lego.com/en-in?consent-modal=show
36. https://www.microsoft.com/en-us/sustainability

37. https://www.patagonia.com/our-footprint/
38. https://www.tcs.com/what-we-do/services/sustainability-services
39. https://hul-performance-highlights.hul.co.in/performance-highlights-fy-2021-2022/environmental.html
40. https://www.itcportal.com/sustainability/index.aspx
41. https://www.tesla.com/en_eu/impact/environment
42. https://www.coca-colacompany.com/sustainability
43. https://www.interface.com/US/en-US/sustainability/sustainability-overview.html
44. https://about.ikea.com/en/sustainability
45. http://patanjaliayurved.org/csr-policy.html
46. https://www.nestle.in/csv/water-and-environmental-sustainability
47. https://corporate.walmart.com/purpose/sustainability
48. https://www2.hm.com/en_in/sustainability-at-hm.html
49. https://www.intel.com/
50. https://www.interface.com/US/en-US/sustainability/sustainability-overview.html
51. https://www.bcorporation.net/en-us/certification/
52. https://www.gsk.com/en-gb/responsibility/environment/
53. https://www.siemens.com/
54. https://www.tatasustainability.com/
55. https://www.novartis.com/esg/environmental-sustainability
56. https://www.starbucks.com/responsibility/planet/
57. https://www.mca.gov.in/Ministry/pdf/InvitationOfPublicCommentsHLC_18012019.pdf

    The Comprehensive Guide On CSR For Consultant

58. https://www.infosys.com/infosys-foundation.html
59. https://www.mahindra.com/rise
60. https://www.wiprofoundation.org/
61. https://www.hzlindia.com/csr/
62. https://www.adanifoundation.org/
63. https://drreddysfoundation.org/
64. https://www.larsentoubro.com/
65. https://www.huf.co.in/en/
66. https://www.hdfcbank.com/personal/about-us/overview/who-we-are
67. https://marico.com/
68. https://www.ambujacement.com/
69. https://www.jsw.in/foundation
70. https://www.titancompany.in/
71. https://www.godrej.com/godrejandboyce/pirojsha-godrej-foundation
72. https://www.adityabirla.com/
73. http://www.wockhardt.com/
74. https://www.apple.com/in/environment/
75. https://sustainability.aboutamazon.com/
76. https://corporate.exxonmobil.com/
77. https://www.ford.com/
78. https://www.microsoft.com/en-in
79. https://www.interface.com/US/en-US/sustainability/sustainability-overview.html
80. https://www.salesforce.com/in/
81. https://www.coca-colacompany.com/sustainability/water-stewardship/replenish-africa-initiative
82. https://www.tatatrusts.org/topics/tata-water-mission
83. https://www.mca.gov.in/Ministry/actsbills/pdf/Societies_Registration_Act_1860.pdf

 The Comprehensive Guide On CSR For Consultant

84. https://dllromsd.org/LAW_WEB/act3.pdf
85. https://www.mca.gov.in/Ministry/pdf/CompaniesAct2013.pdf
86. https://sbi.co.in/corporate/AR1920/download_center/english/11-4.10-Corporate%20Social%20Responsibility.pdf
87. https://www.hdfc.com/
88. https://www.airtel.in/about-bharti/about-bharti-airtel/
89. https://www.soulace.in/
90. https://consultivo.in/social-sustainability/corporate-social-responsibility-csr/
91. https://csrbox.org/Impact-Advisory/
92. https://www.pwc.in/
93. https://corpbiz.io/ngo-csr-consultancy
94. https://www.bain.com/
95. https://wachsstrategies.com/
96. https://www.fticonsulting-asia.com/

www.ingramcontent.com/pod-product-compliance
Lightning Source LLC
Chambersburg PA
CBHW070917260726
48661CB00003B/742